The Changing Face of the Fire Service:
A Handbook on Women in Firefighting

Prepared by:

Women in the Fire Service
P.O. Box 5446
Madison, WI 53705
608/233-4768

Researchers/Writers:

Dee S. Armstrong
Brenda Berkman
Terese M. Floren
Linda F. Willing

Disclaimer

THE U.S. FIRE ADMINISTRATION AND WOMEN'S ISSUES

In August of 1979, the U.S. Fire Administration (USFA) convened a "Women in the Fire Service" seminar in College Park, Maryland. This seminar brought together a group of fire service leaders and others to discuss the relatively new phenomenon of women -- perhaps 300 nationwide -- employed as firefighters. Today, women in firefighting positions have increased to approximately 3,000.

As a result of the symposium, the USFA in 1980 published a document called The *Role of Women in the Fire Service.* The publication summarized the issues discussed at the seminar, presented the participants' recommendations and personal insights into many aspects of women's work in the fire service, and described existing initiatives and resources in the area.

Also in 1980, two manuals were developed as a result of seminar participants' recommendations. The first, a joint project of the USFA and the International Association of Fire Fighters, was a resource directory identifying fire departments that employed women and/or had experience with recruiting, testing and training women firefighters. The second manual, *Personnel Management Handbook: Managing the Entry of Women and Minorities,* focused on legal issues, management commitment, recruitment strategies, and the development of entry-level physical fitness standards.

The fire service has experienced many changes in the twelve years since the publication of these documents. Some important issues have been resolved, and new ones have moved into the limelight. The number of women working as career firefighters and officers has increased tenfold; some fire departments' suppression forces are now ten percent female or even higher. Several fire departments have employed women firefighters for almost two decades, and many have promoted women upwards through the ranks to district chief and battalion chief positions. Other departments, even large ones, have yet to hire their first women firefighters.

The challenge in the development of this document has been to provide a resource useful to all fire service personnel. This new handbook, *The Changing Face of the Fire Service,* updates the information contained in all three of the earlier documents and reflects a decade of experience and progress.

I encourage you to read the handbook from start to finish. The topics reflect important issues of our day. The handbook will serve as a valuable resource to anyone seeking guidance in areas affecting the integration of women into fire suppression positions.

We all benefit from a fire service that is inclusive of women at all levels. USFA is committed to promoting an environment where women and men can work harmoniously and productively together to protect our communities. I believe the information contained in this handbook is an important step in achieving that goal.

Olin L. Greene
U.S. Fire Administrator

TABLE OF CONTENTS

Why a manual on women firefighters?

Fire may know no gender, but people do, and it is the fire chief's job to manage people much more often than he or she manages fire. Women and minorities are forming an increasing part of the workforce and the labor pool. Fire service managers in the 1990's have an opportunity to attract the best of that pool by creating a work environment that welcomes the participation of all. Facilitating teamwork in a culturally diverse fire department may be the biggest challenge currently facing fire chiefs. Those who are not prepared to manage a diverse workforce may find that the workforce is managing them instead.

It's not enough just to say, "We'll hire anyone who meets our standards." How are those standards set? Can they be justified? What happens to someone who "meets the standards" but faces a barrier of hostility and opposition from co-workers? What support systems are available for workers who are not part of the dominant group? Simply having policies in place that appear to be neutral, or are

applied equally to everyone, does not necessarily create equal opportunity. Altering the identity of people in a fundamentally unaltered workplace can leave the door open to friction, miscommunication, and a host of inequities that can result in poor performance and a loss of teamwork.

Despite repeated claims that "We don't want to reinvent the wheel," fire service leaders continue to ignore the critical lessons learned in other fire departments over the years. The ideas and resources in this handbook have been drawn from the experiences of fire departments across the country over nearly two decades of women's involvement in career-level fire suppression. They are offered to all those in the fire service, career and volunteer alike, who wish to see a smooth transition to a gender-integrated workforce, and to be a progressive and pro-active part of thechanging face of the fire service. This book is particularly created for, and dedicated to, fire service managers. The chances for that smooth transition ride on their commitment and hard work.

How to use this handbook

The purpose of this document is to help the fire service manager cope with the firefighting workforce as it changes from an all-male environment to one that includeswomen. The handbookoffers guidance and suggestions from people who have experience and expertise in the areas that affect women's integration into fire suppression ranks. It takes a wide perspective, and, as is appropriate with personnel issues in general, offers choicesand options more often than single "right" answers. The authors hope they have created a guide that, in calling on a wide range of resources, can be useful to a wide range of needs, whether that means one woman firefighter wanting to know where she can find a pair of gloves that fits, or a fire department management team seeking an overview on gender integration issues.

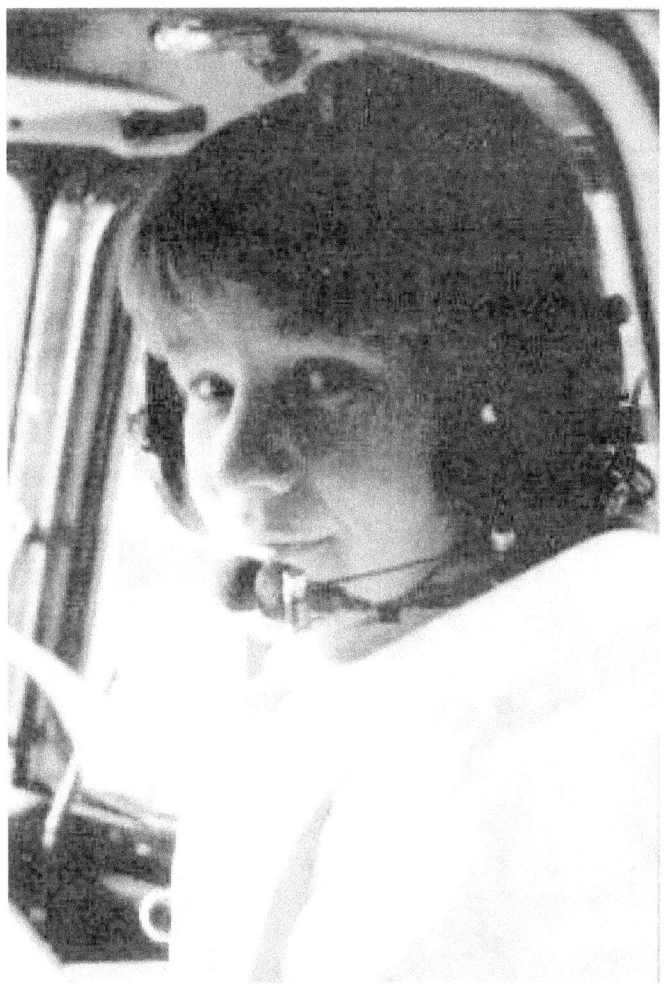

The Changing Face of the Fire Service attempts to answer the most frequently asked questions about women in firefighting. It focuses specifically on the issues of recruitment, entry-level physical testing, firefighter training, maternity and reproductive safety, hair-length standards, fire station facilities, sexual harassment, cultural diversity training, ongoing support, and protective gear and uniforms. Each issue is explored from both legal and practical standpoints. In the area of policy development, sample language has been included in some cases. Where it was felt that existing policies would quickly become outdated, guidelines for policy development have been substituted for specific language. Sources for obtaining up-to-date samples of policies are included in the "Resources" section of the manual.

The handbook has two specific limitations. Although it has been reviewed by several attorneys, it does not consider the requirements of most state or local laws and regulations. And, while the information in the manual was up-to-date as of mid-1992, many of the legal issues in question are subject to change. The handbook is not intended to be a substitute for qualified legal advice, which should be sought before implementing policies that have legal implications.

The Changing Face of the Fire Service was prepared under contract to FEMA by Women in the Fire Service. It was made possible only with the assistance of dozens of individuals from fire departments and other agencies throughout the country who provided information and shared their valuable insights. For this assistance, the researchers and writers of this document offer their sincere thanks.

As we enter the last decade of the twentieth century, more than 3,000 women are at work in career-level fire suppression positions on 650-plus fire departments in the United States, with hundreds of counterparts in Canada, France, Denmark, the United Kingdom, Australia, and New Zealand. Thousands of other women work for fire departments in EMS, fire prevention, arson investigation, communications, and public education. Many women firefighters have been promoted to the rank of engineer, lieutenant, or captain, and the numbers of battalion chiefs, division chiefs, and chiefs of department continue to increase. At least twenty women in the U.S. have made their way up through the ranks to the level of division chief, battalion chief, or district chief. Some large cities and many smaller towns and fire districts have women serving on their civilian boards of fire commissioners.

While women first became career firefighters just two decades ago, they have served as volunteer firefighters for more than a century. Among the volunteer and paid-on-call fire and EMS forces in the United States are perhaps thirty thousand women firefighters, and thousands more EMT's and paramedics. A large number are officers, including many chiefs of department. Many of these women are dedicated and skilled firefighters and rescue workers with a lifelong commitment to their volunteer departments; others, equally skilled, plan to move on to career positions in the fire service.

How many women are volunteer firefighters?

Exact numbers on volunteer firefighters are hard to obtain, as the population is highly mobile, and often no system is in place to track it. Most states do not keep accurate or centralized records of firefighters: some do not even know how many firefighters there are in the state. Even states that do keep records often do not have a system that can draw out race or gender statistics on either career or volunteer firefighters. In 1986, however, the National Volunteer Fire Council commissioned a study of volunteer firefighters in five states.[1] Of the volunteer firefighters surveyed, 3.3% were women. Applying this percentage to the country at large provides an estimated number of 42,000 women volunteer firefighters.

A few states offer statistics that are much more specific. In Virginia, the Department of Fire Programs publishes an annual profile of minority and women's participation in their Fire Services Training. Their 1991 report indicated that 1,249 of the state's 25,857 fire personnel (assuming all firefighters participated in the training system at some point during the year) were women, or 4.8% of all firefighter participants in the system. Significantly, that number had increased more than 9% from just the preceding year. The report indicated that women represented 2.9% of career firefighters and 5.7% of volunteers. (These are suppression-only numbers: women in strictly EMS roles constituted 38% of participants in the training system.)

Extrapolating from these numbers provides a second estimate of the national picture. The percentage of career firefighters in Virginia who are women is slightly more than twice the national average. If we assume this to be true for volunteers as well, we can calculate an estimate of 36,300 women volunteer firefighters in the U.S., as well as a much greater number of women who are volunteer EMT's and paramedics.

In Vermont, one of the least populous states, 90 women are volunteer firefighters (including two lieutenantsand one captain), and three women are career firefighters. At the other end of the spectrum is California, which has more than 600 women in career fire suppression roles and an unknown but certainly much greater number who are seasonal or volunteer firefighters.

Numbers and ranks

The national women firefighters' organization, Women in the Fire Service, conducts periodic surveys of women firefighters, focusing on career-level women in suppression roles. The most recent one, in 1990, gathered information from 425 women nationwide, of whom 356 were career firefighters or officers. The following profile is taken from the data from that survey.

Fire departments that employ women range in size from small combination departments with only a few full-time firefighters, to New York City with its nearly 12,000 line personnel. Percentages of women on those fire departments range from minuscule (one or two women out of several hundred firefighters) to one department whose only paid person in suppression is a woman lieutenant. As of 1992, the suppression forces of several severable fire departments were over or approaching ten percent women: Madison, Wisconsin

(31 women out of 260 total firefighters); Montgomery County, Maryland (92 of 900); San Diego (76 of 900). Some smaller departments had even higher percentages: Boulder, Colorado (14 of 72); Davis, California (5 of 40); and Littleton, Colorado (10 of 80). The California Department of Forestry and Fire Protection, which has 3,412 full-time personnel who perform both wildland and structural firefighting, employs some 229 women, the largest number of any single employer. At least 84% of women firefighters on departments represented in the survey had been hired without recourse to court orders or consent decrees.

Women are not only firefighters, but officers as well. Among the career ranks in the U.S., there are at least 119 women driver/engineers (on departments where that is a promoted position), four sergeants, 72 lieutenants, 54 captains, and 21 battalion or division chiefs. Several women are assistant chiefs. At least three women are chiefs of paid or combination fire departments. In addition, some departments with officer ranks in their EMS division have women who are senior paramedics, paramedic lieutenants, or battalion chiefs in charge of EMS.

Who are women firefighters?

Women in the career fire service come from a wide range of occupational backgrounds, both traditional and non-traditional. Many were teachers or worked in health care (as nurses, EMT's or paramedics); others were in the military, worked as secretaries, or were full-time mothers prior to becoming firefighters. Still other women firefighters come from fields as diverse as carpentry, landscaping, auto parts, and photography. Most women who are career firefighters have college degrees, either two-year or four-year, and several hold masters' or doctoral degrees.*

An estimated eleven percent of women career firefighters are African-American. Another four percent are Hispanic, one percent Asian, Native American, or other women of color, and 84% are Caucasian. Forty-one percent are married, 20% are unmarried but in committed relationships with a partner; 39% are single. Thirty-four percent of women firefighters are mothers. Women firefighters at the career level range in age from 21 to their early 50's; women volunteer and cadet firefighters may be as young as 14 or as old as 70.

These statistics demonstrate one thing most clearly: there is no "typical" woman firefighter. Women in the fire service exhibit at least as much collective diversity as fire service men, in terms of age, ethnicity, size, work history, and personal backgrounds. A woman firefighter may be a 23-year-old, 5'4", 120-pound former factory worker who's married and has two children, or a 41-year-old, 5'11", 185-pound former concert violinist who lives alone and climbs mountains in her off-duty time. What these women have in common is their dedication to their work as firefighters, and the intensity of their commitment to the fire service.

Many people feel that inequality based on sex is natural, functional, and in large measure unalterable. Explanations for inequality that would be unacceptable in the context of race pass without substantial objection in the context of gender: for example, the discredited notion that a woman's hormones would interfere with her ability to be President. Perhaps as a result of these miscon-ceptions, many employers have rejected efforts to promote gender equality because they seemed to be economically unworkable. This attitude has produced long-term costs that are less visible but dramatic: lost talent, heightened turnover, and diminished productivity, to name but a few.

Some employers have been reluctant to accommodate employee differences or to reformulate jobs in ways that would allow previously excluded people to perform as well as workers who have been accepted under current standards. (An example would be the purchase of vehicles equipped with power steering, to reduce the amount of upper body strength that drivers need.) This reluctance forms a continuing barrier to equal employment opportunity. Women and minorities, once admitted to an occupation, are expected to accommodate themselves to existing institutional norms; those norms are not expected to change. Providing special treatment for previously excluded groups is criticized. However, "special treatment" is a highly subjective term.

The difference between the way "accommodation" provisions in the Americans with Disabilities Act and "special treatment" for women workers are regarded shows this subjectivity very clearly. We applaud moves that make reasonable accommodation for people with disabilities. A slight modification of access can allow a person in a wheelchair to do a job they had been prevented from doing even though they were fully capable of its demands. We have come to recognize that accommodation implies differences, not inferiority.

This extends to differences that are not considered disabilities. Fire instructors allow left-handed firefighters to tie knots left-handed even though that isn't the way most people do it; a difference in "handedness" is an acceptable variation. All firefighters applaud advances in technology that allow them to do their jobs more easily: lighter-weight SCBA bottles, pull-out steps or hydraulic racks for mounting ladders on vehicles, wheeled carts for smoke ejectors, pulleys that provide mechanical advantage on ladder halyards. However, the standards for determining when a variation among people is not acceptable, or when a technological advance means that the job is being "made easier" in some unacceptable way, are set from the highly subjective perspective of the dominant group.

When is a firefighter "too short to do the job," and when has a ladder simply been placed too high on the vehicle? Techniques and technology that make the job easier are usually hailed as advances. But women's presence on the job often makes such advances suspect: the job is being "made easier" in some illegitimate way so that "unqualified" women can be hired. The change becomes an unacceptable form of "special treatment," even though the result is that the job is made safer and easier for everyone. The fact is that any change that makes the job less physically demanding will both save firefighters' health, safety, and lives, and make the job just a little more possible for someone who wasn't capable of it before.

Members of a dominant group within any institution tend to view those who are not members of that group with skepticism. One way this dynamic affects women firefighters is in a strong insistence that they prove themselves by doing everything the hard way. Differences in technique are deemed unacceptable. But the "special treatment" that is criticized in the fire service is defined from a male reference point. If the workplace had been designed for and dominated by women for over a century, what accommodations to male cultural norms would we now regard as "special treatment?"

It is not enough just to allow members of traditionally excluded groups to apply for firefighter positions. We must re-think the direction of the fire service. What is "equality of opportunity?" It is not a guarantee, but it must be more than a mere possibility. Changes in firefighter stereotypes will come slowly. But change must come if the fire service is to deliver optimum service to the community. We all have the opportunity to be part of the future by examining our hiring practices, our organizational policies and our own assumptions from a fresh and critical perspective.

Notes:

[1] Perkins, Kenneth B., "Volunteer Fire Fighters in the United States: A Sociological Profile of America's Bravest," National Volunteer Fire Council, 1987.

[2] Willing, Linda F., "Origins: Who We Are and Where We Came From," *WFS Quarterly*, pp. 4-8.

The true measure of the success of a recruiting drive aimed at women is found...in the number of women who are on the job two or three years later as skilled and productive firefighters.

Recruitment is usually the focal point of a fire department's efforts to increase the number of women firefighters on the job. In the traditional view, it consists of the effort that is spent, shortly before an application period opens or an entry-level test is given, to get candidates to apply for job openings. This section of the handbook outlines the basic elements of such recruitment programs and presents some insights from programs that are already in place in various parts of the U.S.

A more comprehensive view of recruitment, however, shows that much more is involved in increasing the number of women on the job. This section, therefore, also attempts to expand current approaches to recruitment in two important directions: (1) by emphasizing the importance of laying the necessary groundwork within the department before ever seeking applications from women candidates; and (2) by recognizing the recruitment impact of many fire department activities.

Before recruitment begins: laying the groundwork

The skills and dedication of the people working on recruitment, the creativity that goes into designing the program, and the support - financial, logistical, and verbal - given to the effort by top management, all play important parts in the success of your department's recruitment effort. However, the entire operation can be negated or undermined from within the department. Your recruiters' message will be that your fire department wants women firefighters. If other aspects of your department give out a conflicting message, or if the department is unprepared for the change to a two-gender workforce, much of your recruitment effort will go for nothing.

An example will help illustrate this. Suppose that, as a result of recruitment, hundreds of women apply for firefighting positions and show up to take the test. Does that mean your recruitment efforts have been successful? Of course, the answer is, "Not necessarily." Yet many fire departments and recruiters emphasize sheer numbers of women applicants as an indicator of effective recruiting. Giving greater consideration to what you're really trying to accomplish will show this to be short-sighted. The true measure of the success of a recruiting drive aimed at women is found much farther down the road, in the number of women who are on the job two or three years later as skilled and productive firefighters.

That involves factors that aren't the job of the recruiters - which is just the point. The out-front recruitment effort is just the tip of the iceberg. Its long-range success hinges as much, or more, on work that must be done elsewhere in the department. If the recruiter's job is to make it known that the department sincerely wants women to work there as firefighters, it is the fire chief's job to make sure that's true.

A pre-recruitment checklist

This checklist covers some of the basic preparatory steps that fire department managers can take to make the recruitment of women firefighters more effective, and to insure the retention of those women who are recruited.

The application and testing processes:

❏ For volunteer fire departments that do not give an entry-level test, has the application process - including both formal and informal elements - been reviewed to insure that it is welcoming of, and accessible to, women?

❏ For departments that give an entry-level test, has the testing procedure been established in full detail?

❏ Will candidates be informed about the components of the process, the elements of the physical test, how the tests will be administered, how each component will be scored, what constitutes a passing score, and how the final ranking on the eligible list (or determination of who is hired) will be made?

❏ If your entry-level physical test is one that women pass at a significantly lower rate than men, has it been locally validated, based on a job analysis conducted by an outside expert?

❏ Can the scoring system used on the test be justified as accurately predicting job performance? Do those who score lower on the test make less competent firefighters, after they have been trained, than those who scored higher?

❏ Has the pass/fail point been set by testing a random sample of firefighters already on the job?

❏ Have trainable items on the test been kept to a minimum?

❏ Have those who will be administering the test been instructed on what variations in technique on the test events will be permitted? Will a variety of safe and effective techniques for completing the events be permitted and demonstrated?

❏ If candidates will wear items of protective gear during the test, are these available in sizes to fit all candidates?

❏ If test practice sessions will be held, are instructors or assistants for the practice sessions familiar with techniques that may be more effective for women and/or shorter individuals? Has there been good coordination between these instructors and the people who will administer the test, so that all permissible techniques are demonstrated, and accurate information about times, items and evolutions used on the test, and the sequence of events, will be given out?

[See section on entry-level physical testing for more information.]

Policy development and review

❏ Have an anti-harassment policy and complaint procedure been implemented and publicized? Do personnel understand the procedure for filing sexual harassment complaints, and do they trust the effectiveness and confidentiality of the process?

❏ Has cultural diversity and/or anti-harassment training been conducted for all personnel and been incorporated into the recruit school curriculum?

❏ Does the department have policies in place regarding maternity and fire department marriages?

❏ Have policies regarding hair length and personal grooming been reviewed?

[See sections on sexual harassment, cultural diversity training, and policy development.]

A pre-recruitment checklist, continued

Recruit training

❑   Does the training staff have a positive attitude towards training women, and are they knowledgeable about cultural barriers that may exist for women coming onto the job? Are all instructors available after hours for additional assistance if requested?

❑   Have interested women been detailed to the training academy as instructors or assistants?

❑   Has the training curriculum been examined to see if it implicitly expects new recruits to have particular skills or backgrounds? Are there ways to impart this information or attitudes, or these skills, to those who don't possess them?

❑   Has the training staff been clearly instructed on how to handle difficulties faced by new recruits: how to document any problems, how to assist in the learning process?

❑   Are performance criteria for training clear and will they be clearly explained and/ or distributed to all recruits so they know what is expected of them? Are recruits taught skills (hands-on, not just observation) before being evaluated on them, and will recruits be informed at what point evaluations will take place? Do training dynamics give recruits a stake in each other's success and promote cooperation and teamwork?

[See section on training.]

Gear and facilities

❑   Are uniforms available in women's cuts and sizes?

❑   Will properly fitting protective gear be available to all firefighter recruits from the time it is first needed in training?

❑   Does the training center offer adequate, private facilities for both sexes: not just restrooms, but changing areas and showers?

❑   Are all fire stations adequate in their facilities for a workforce that includes people of both sexes? If not, has a plan or timetable been implemented to make them adequate?

❑   Will women be permitted to work at any station, regardless of whether the station has "facilities for women?"

[See sections on facilities and uniforms /protective gear.]

What is clear from this checklist is that most of the issues that arise when a fire department becomes gender-integrated should be considered before recruitment ever begins. A fire department that has little management commitment to hiring women, an unvalidated speed-to-completion entry-level physical test, haphazard training, hair-length standards designed for men, and a"We'll deal with it when it happens" approach to maternity leave, is showing that it truly does *not* care what happens to any women who might be hired. As one fire chief said, "It was hard to get women to come on the job, because the men didn't want them." Potential recruits are unlikely to be motivated to join a department on which they are unwelcome. In addition, women firefighters who have been asked to be involved in the recruitment effort may be reluctant to do so, since they know the problems that any women coming on the job will face. It will be difficult, both practically and ethically, to recruit women into such a work environment.

A fire department's real recruiting effort, like its public education work, goes on all year round. Volunteer fire departments, many of which are in constant need of new members, already know this. Whether consciously or otherwise, a fire department recruits new members and gives out information about itself all the time. Being aware of the recruitment potential of everything the department does will be both efficient and productive.

The women and the people of color currently on the department should be included in all of the department's public activities. Firefighters "self-recruit" - that is, attract more people just like them - because of their visibility. If the only firefighters who are visible are white men, that's largely who will be recruited. Make sure that every time the department is in the public eye, and especially when it is being televised, the diversity of your department is visible.

Maintain a consistent commitment to non-sexist attitudes and language. Using and encouraging gender-neutral language is an important statement. For example, does your local newspaper still refer to firefighters - of either sex-as "firemen?" A letter to the editor from the fire chief will not only help get the practice stopped, but will in itself demonstrate your department's commitment to fairness and diversity.

In addition to maximizing the department's public appearances to benefit recruitment, specific steps, such as those described below, can be taken to encourage potential candidates to consider, and prepare for, firefighter jobs. These are all long-term efforts that will only produce results over the course of years. If you do not undertake them now, however, your department will continue to be in the position of having to depend on short-term recruitment to attract women who have never before considered becoming firefighters. The more information about the job you can provide to possible future employees, the easier it will be to locate and attract good, qualified personnel.

Vocational counselors

Develop and maintain a good relationship with high school guidance counselors and with career-placement personnel and vocational counselors at the colleges and universities in your area. These people can be where you can not: in contact with young people who are making career decisions. Are the counselors in your colleges and high schools currently encouraging young women to consider fire service careers? It's most likely that they are not. Make sure they are aware of your department's interest in hiring women firefighters, and provide supplies of literature, videotapes, and other information

for them. Invite them to orientation sessions, or consider holding special sessions for counselors to discuss the ways you and they can work together.

Introductory programs

These efforts can be a productive way to recruit new firefighters who, by the time they complete the program, will be familiar with your department's operations and may possess basic firefighter training as well. Such programs include Explorer posts (open to young men and women 14-21 years of age), high-school or vocational school "fire cadet" programs, ride-alongs for community members, resident student programs (where students bunk in the stations and, after completing basic firefighter training, ride along on calls). All of these can be effective recruiting tools. They will be effective at recruiting women only if you make sure they are inclusive of women, and that women are specifically sought as participants.

Extended contact with potential candidates

Many elements of a recruiting program can easily be kept operating at all times. Periodic open houses, practice test sessions, and orientation sessions for potential candidates can motivate candidates far in advance of test dates, allowing more time for them to develop their strengths and skills or to seek relevant education.

The fire department, or the personnel department, should accept job interest cards at all times, even when a test is not planned. To keep the data base current these could have a one- or two-year expiration date, at which time a card would be sent out to verify the individual's continued interest. A verification card could also be sent out if a hiring process opens up and the person does not apply; those who do should automatically be kept on the list for another year or two years.

> These are all long-term efforts that will only produce results over the course of years. If you do not undertake them now, however, your department will continue to be in the position of having to depend on short-term recruitment to attract women who have never before considered becoming firefighters.

Some fire departments do a general publicity and recruitment drive when their test is announced, or - for volunteer departments - when more firefighters are needed. The purpose of the drive is to get job information to as many potential applicants as possible. Many departments, however, have found that large numbers of well-qualified white male applicants will apply for firefighter positions even if the openings are not publicized at all. This can be because of a strong family-and-friends tradition within the department, or more generally because white men as a group are well aware that the fire service is a career option for them. The department decides then to focus its limited resources of time and money on a recruitment effort that targets those groups that are underrepresented in the workforce or applicant pool. No one is discouraged from applying for the job, but the tendency for the existing workforce to "self-recruit" is counterbalanced by publicity aimed at those who might not otherwise apply.

Setting up a recruitment program involves planning, commitment, creativity, and may require the coordinated work of a number of people from different departments. A recruiting drive can be as basic as one person with a slide show working for three weeks, or as complex as a fully staffed division operating over the course of a year. The exact details of the effort will vary considerably depending on several factors, including:

The size and resources of the particular fire department;

The goals of the recruitment program;

The amount of time available for recruiting;

The make-up of the community and the nature of the labor pool; and

The creativity and enthusiasm of the people involved in designing and delivering the program.

Whatever its dimensions, a well-designed and carefully organized recruitment program will always bring greater success than one that is haphazard or based on misconceptions. Certain key elements will be present, in one form or another, in any effective and successful recruitment drive. These elements, all of which are discussed in detail on the following pages, include:

Management support;

Careful recruitment team selection;

Realistic schedule design;

Recruitment materials aimed at the target group;

Publicity within the community;

Effective use of the media; and

Orientation sessions and open houses.

Management support

The top individuals in the fire department must be firmly behind the recruitment and integration of women into the department. All aspects of the recruiting effort must reflect management's sincere commitment not only to bring women firefighters onto the department but to support a diverse fire service workforce. Management can support recruitment in the following crucial ways:

● Obtaining the funding necessary to make the program a success, either by making it a priority within the department's budget, seeking additional city funds (such as from the personnel department or any agencies concerned with equal employment opportunity), or obtaining donations from sources in the community;

● Making other departmental resources available to maximize the allocated money (reassigning fire personnel and support staff, vehicles, office space), and borrowing or getting access to other city resources (prior applicant lists for non-traditional jobs from the personnel department, photographic and darkroom work from the police department, audiovisual equipment and expertise, etc.)

● Working with other city departments as needed to obtain adequate lead time (the time between the announcement of the test date and the actual date, during which most short-term recruitment will be done) and to settle any other jurisdictional or political problems. For example, restrictions on how applications are given out can make it more difficult for women to obtain them. One fire department required applicants to come to its headquarters in person to pick up the necessary forms; the headquarters building was in a high-crime area, and numerous assaults on women had taken place in nearby parking garages. Making the forms easier to get will make it more likely that women will apply. Since this is often an inter-jurisdictional matter (that is, not one that the fire department itself controls), management's influence can be crucial in making the necessary changes.

● Demonstrating leadership by representing the program positively to elected officials in order to obtain their support, and by making public statements, particularly in the media, in support of the recruitment effort and of hiring women and minorities.

Consider carefully before you emphasize numbers as a measure of the success of your recruitment effort, either publicly or within thedepartment. If potentialcandidates and incumbent firefighters perceive (correctly or incorrectly) that management just wants to hire women to get numbers to fill a hiring goal, the sincerity and effectiveness of your recruitment will be severely undermined. The message that "We want to hire ten women" implies two things, and both of them are negative: (1) that you will hire ten women just to hire women, even if not all of them are qualified, and (2) that if more than ten qualified women apply, you will not hire all of them. It also can make your recruitment drive appear to have been a failure if you "only" end up hiring nine women. Instead, consider making a positive "goals statement" that emphasizes your commitment to diversify your firefighting workforce and to support its diversity in meaningful ways. For example, one fire department said in its advertisements:

> We are looking for professionals who want to be part of a progressive, innovative fire department. Our goal is to have a workforce that reflects the diversity of our community. Women and people of color are especially encouraged to apply.

Recruitment team selection

Select members of your recruitment team who have the qualities that will make them effective at their job. Too often, recruiters are chosen for reasons that make sense from a limited perspective but are irrelevant or unproductive for the purposes of recruitment, such as:

S/he has "always been in charge" of recruiting;

S/he is injured or pregnant and needs a light-duty assignment, or for some other reason seeks a 40-hour schedule; or

S/he belongs to the group that is being targeted for recruitment (e.g., people of color, women, paramedics, college graduates.)

All of these practices can create problems. People who have "always" done recruitment will usually continue to produce the kind of candidates they have produced in the past. If these are not the people you're looking for, it could be time to make a staffing change. A firefighter who has been removed from the line due to injury or some other reason is usually chosen for the department's convenience, notbecause of the individual's qualifications to be a good recruiter. (If such people should happen to possess the skills you need, however, they certainly should not be overlooked.) And although it is less obvious, the same is true of firefighters or officers who belong to the targeted group.

Women currently on the job who have an interest in recruitment should be used in recruiting drives. The critical point at which to do so, however, is at orientation sessions and other public-contact points where women who are potential candidates will want to hear from, and ask questions of, women who are currently on the job. But a good speaker or advocate is not necessarily a good program manager. Often, women firefighters who are assigned to coordinate recruiting programs have few credentials for the job and, in some cases, little interest in it. By all means, take advantage of the abilities and interest of those women who do want to be involved; they're highly valuable to you. Don't, however, assume that someone will be a good recruit-program coordinator just because she's a woman and a firefighter.

Identify the skills you will need on the recruiting team before you select its members. Interest in, and commitment to, the recruiting drive, are prerequisites: no one should be chosen who does not want to be involved. Useful skills and traits include:

Education and experience in marketing and public relations;

Graphic arts skills;

Writing skills;

Public speaking skills;

Interpersonal skills for speaking with individual candidates;

Computer literacy: word processing, data-base management, and desktop publishing;

Bilingual ability, if your community is ethnically diverse; and

Ties with target groups.

No individual will possess all of these traits. Diversity on the team is important for that reason, as well as to provide the flexibility and the range of creativity that will permit a variety of approaches. The recruitment coordinator and other top recruiters must also have good organizational skills and be able to work well together in a concerted effort.

The people involved in the recruitment program may come from various areas both within and outside the department. They may include firefighters, non-suppression fire department personnel, support staff from the department or loaned from elsewhere in the city or county, community volunteers, and members of other local fire departments (if your own department has no women, or if you are participating in joint recruitment). Smaller departments may also need to borrow the clerical or computer services of other city departments.

Whatever the size of your department, tap the resources you already possess. A recruitment program is very similar to a public education effort, and your public education staff should be a gold mine of assistance. They know how to scale a "motivational" message to a target audience and present that message in a way the audience will understand and respond to. Similarly, your public information officer has skills and contacts that can be very useful in designing and distributing press releases and in assuring media coverage of recruitment events.

An Incident Command approach to recruitment team function

This outline identifies the tasks or functions that must be included in a recruitment program. As with an emergency incident command system, the important part is not the title or rank of an individual in any particular role, but that the function is assigned to someone who will be responsible for its accomplishment.

Upper management representative:
- Oversees the recruitment effort and pre-training program
- Authorizes necessary expenditures
- Reports on the programs to chief of department, city manager, council, *etc.*
- Attends Steering Committee meetings to get the committee's input

Recruitment Coordinator:
- Develops and implements the overall recruitment plan
- Under direction of upper management, develops recruiting/pre-training staff.
- Coordinates overall program: materials development, scheduling, roles of staff members
- Maintains records of all recruitment activities, including candidate data base
- Provides progress reports to upper management
- Sits on Steering Committee

Human Resources Manager:
- Provides orientation and sensitivity training to recruitmenffpre-training staff
- Works with media coordinator in developing outreach materials
- Sits on Steering Committee

Media Coordinator:
- Develops all advertising, press releases, PSA's, flyers, brochures, and posters to be used in the recruitment effort
- Schedules press interviews, coverage of events, advertising placement
- Coordinates production and distribution of outreach materials
- Works with support staff during photographing/videotaping for the recruitment and pre-training efforts
- Sits on Steering Committee

The **Steering Committee** includes a representative of upper management, the Recruitment Coordinator, the Human Resources Manager, the Media Coordinator, and the 40-hour recruitment/pre-training staff members. It also includes representatives from community organizations used for recruitment and pre-training classes. The Committee meets weekly to coordinate all recruitment and pre-training activities and resources.

Pre-training Coordinators:
- Should have background in exercise physiology and/or physical test preparation, particularly dealing with women candidates
- Develop and implement written and physical pre-training programs for candidates
- Work with community organizations to schedule and conduct pre-training classes

Recruiters:
- Include at least one representative (preferably two) from each group that is being targeted in the recruitment effort, and others as needed
- Staff the intake office
- Contact community groups to schedule orientations and drop off materials
- Attend orientations, special events, and pre-training sessions
- Perform canvassing and retention
- Provide data entry and record-keeping as needed

Support Services:
- Attend to supplies and housekeeping needs
- Coordinate vehicles for recruitment purposes
- Coordinate audiovisual equipment needs
- Design and care for portable recruitment booth
- Do all departmental paperwork: log work time, send in payroll records, log vehicle mileage, gas and maintenance, etc.

[Thanks to the San Francisco Fire Department for providing the model on which this outline is based.]

Community volunteers can be used to distribute literature and to make contacts with various community groups. Small departments with limited budgets may be able to find volunteers within the community who will donate their professional skills to design a brochure. Local businesses may be willing to donate all or part of the cost of printing literature and posters for the recruitment drive. Cable access television channels may provide video production equipment and editing facilities. Fitness centers and gyms may be willing to offer discount memberships to firefighter applicants who are preparing for the testing process.

Realistic schedule design

The coordinator of your recruitment effort will be responsible for its general design. This means determining which fire department and community resources are to be used, what kind of media publicity will be required, what the priority markets are, and what is to be done when.

The first step is to make a checklist of all of the tasks to be completed. This will include items such as:

Develop the budget for the recruitment program, based on total allotted funds and any donated resources;

Identify useful community resources, make preliminary contact with key individuals in each group, and get information on when regular meetings or special events are coming up;

Identify career fairs and similar events that are scheduled during the recruitment period

Write and reproduce brochures, test information, and other literature;

Design and print posters and other items requiring graphic art or photography;

Contact the media: write television and radio public service announcements (PSA's), write newspaper advertisements and press releases; arrange for newspaper articles and television news coverage;

Produce videotapes for orientation sessions and test familiarization;

Schedule orientation sessions;

Set up "open house" dates at fire stations;

Set up physical test practice sessions and written test study sessions; and

Get the logistical support you will need: vehicles, office space and supplies, phone lines, answering machine, chairs, audio-visual equipment, security for office use at night, etc.

Next, arrange the tasks into a schedule, time-line, or action plan. This should include when each task must be done, who will do it, how long it will take, what resources will be needed to carry it out, etc. Allow time for

> **Recruitment efforts must be given adequate time in order to be effective, particularly when the recruitment is aimed at women.**

unforeseen delays: don't schedule a major event for the day after needed materials are due from the printer. If bad weather might force cancellation of a specific event, schedule an alternate date in advance.

Recruitment efforts must be given adequate time in order to be effective. This is particularly true when the recruitment is aimed at women. Deciding to become a firefighter is not an easy decision for many women. Issues of self-image and self-confidence, a partner or spouse who may not be supportive or understanding, the demands of family responsibilities, and the risk involved in leaving a job to start something new and uncertain, all frequently arise and can not always be resolved quickly. Women may also need time in which to prepare for the physical test and the physical demands of the job.

If you are limited by decisions made in other departments - for example, Personnel or Civil Service sets the test date, and will only give you a few weeks' notice - many of the above items can be developed in advance and kept on hold until the date is known. For example, you can write a budget, contact community resource people, select your program staff and coordinator, establish a time line to be put into operation once the date is known, and draft your written materials (leaving the date and other undetermined factors blank). You can also make a videotape that explains and demonstrates the testing process, if the test will not be changed before it is administered. Contact friendly reporters with the local papers and television stations to let them know the type of events you'll be offering them a chance to cover once recruitment is underway. Fire departments that can not obtain adequate lead time for an effective recruitment drive must also rely heavily on the ongoing, year-round types of recruitment discussed on page 13.

Recruitment materials aimed at the target group

In designing and writing recruitment materials, remember that you are trying to appeal to a different group of candidates from those who have traditionally applied for the job. Specifically recruiting women to become firefighters is not just a matter of going to places where there are lots of women - even lots of women who are likely candidates - and handing out the usual information about application deadlines and test dates. The content and image of your recruiting message must be different and must address cultural preconceptions that women have about themselves and the job. A brochure or insert specifically aimed at women candidates can be highly effective at delivering your message.

Your primary piece of recruitment literature should be a brochure that presents information about three things: the job of firefighting, your fire department, and the upcoming testing process. The design should be simple enough that your budget can afford to have hundreds or thousands of copies printed and your recruiters will be able to distribute them widely. Some departments have these produced professionally and find the results well worth the investment. However, it is quite possible to put out a high-quality product in-house if no funds are available for outside assistance. A compromise is to have a single sheet or folded flyer produced professionally, that contains information that will not change (about the job and the department), and add or insert into this material produced by the department. Because the cover sheet or exterior of the flyer can be used for several years, you will not have to reinvest in its production, and you can take advantage of volume discounts in the original printing. Your own added material, which contains information about the next test, current salary and benefits, etc., keeps the literature up to date. If the outside of the brochure leaves space for an address and postage, mailing the information to interested candidates will be much easier.

It is important that the text of the brochure explains all aspects of the firefighter's job, including both its demands and its rewards. Women tend to be attracted to firefighting for the same reasons men are: the challenge of a physically demanding job, the rewards of performing a service to the community, and good pay and benefits. All of these should be emphasized.

In writing your brochure, keep in mind how little the average non-firefighter knows about firefighting and about the other work that firefighters perform. Women in particular may have preconceptions about the job that are inaccurate and can either keep them from applying or can lead to problems later on, if the job turns out to be different from what they expected. Do not understate the risks involved in the job, but don't overemphasize them, either. Make it clear that all recruits will be fully trained by the department before ever being put into an emergency situation, and that safety is always a priority. Discuss recruit training: how long it lasts, what the schedule is, what is taught. If your station facilities are designed to accommodate a two-gender workforce, be sure to mention (or show photos of) that.

The recruitment brochure should use gender-neutral language. This not only means saying "he or she" instead of "he," but also substituting terms like "staffing" for "manning," and saying "women," not "females." Little things count, and they contribute in big ways to the impression you make. The overall message should not be that women can manage to perform the job of firefighter and somehow fit into a "man's" job, but that women can

Recruiting women as volunteer firefighters

Volunteer fire departments vary even more widely than career departments in their acceptance of women firefighters. At their best, they welcome anyone who has the abilities and dedication required to be a firefighter. Many women have received strong support from the men on their volunteer fire departments, and many volunteer departments have women lieutenants, captains, and even chiefs. At the other end of the spectrum are volunteer fire departments that adhere rigidly to a "white men only" tradition, excluding others either by specific prohibition or by the force of custom. People of color and women who do manage to join departments of this sort often face multiple barriers such as isolation, lack of training, and overt forms of harassment.

The legality of maintaining a sex-segregated volunteer fire department is a question that must be answered on a case-by-case or state-by-state basis. Some states, such as New Jersey, consider volunteer firefighters to be employees for the purposes of Title VII, which means that sex discrimination against a volunteer firefighter or applicant is illegal. In other states, this is not the case. *[See Appendix 1 for more informatiom.]* The issue is irrelevant, however, to a discussion on recruiting women firefighters, since departments that wish to exclude women are unlikely to spend time actively recruiting them. Women's participation in volunteer fire departments goes back more than a century, and for many (if not most) volunteer departments that have seen their numbers dwindle over the last twenty years, the question is not, "How can we keep women out?" but "How can we get more to join?"

Most of the issues discussed in this section of the manual apply in some form to volunteer departments. The scale, economics, and some of the details differ, but the work and traditions of firefighting are the same. For example:

● Volunteer fire departments often do not have entry-level physical or written tests. However, it is wise to assess the entry process that candidates must go through to become volunteers. Does it contain elements that make it easier for women to be excluded, such as a provision that all applicants must be voted on by existing members?

● Policies and education against sexual harassment are just as important in volunteer departments as they are in career departments; perhaps even more so in states which do not offer employment-discrimination protection to volunteer firefighters.

● Short-term child care during emergency calls will often be a significant issue for women volunteer firefighters and firefighter couples. Creative solutions can both take care of the problem and demonstrate your department's support for women.

● Volunteer fire departments will usually need to rely more on donated resources in order to put together a recruitment effort. The positive side of this is that people in smaller communities served by volunteer departments are much more likely to donate their time and expertise than they would to a large, career-level city department.

Most volunteer fire departments are constantly seeking new members. Giving attention to the above issues, and adapting the suggestions presented throughout this section to your department's specific needs, should make it more likely that your department will be able to attract and retain more women firefighters in the future.

---

and do enjoy productive fire service careers. Photos or drawings in the brochure should include people of both sexes and a variety of ages, sizes, and ethnic backgrounds.

Your literature should include the customary information on pay, hours and benefits, pension system, number of stations the department has, average call load, and so forth. It should also provide clear explanations of the following items:

Details of the application, testing and hiring processes, including age limits for hiring, eyesight requirements, medical examination, and drug screening;

How and where to get application forms;

Whether applications will be available by mail or can be picked up by one person for another;

What to bring if applications must be picked up in person (for example, to prove identity, citizenship, or city residency);

Whether turning in an application early can confer any advantage, such as in the case of a scoring tie, or if a limited number of applications will be given out; and

Whether applications must be filled out on the spot, or may be returned by mail.

Information on the testing process should clearly state which test will be given first, and when and how applicants will be notified whether they have passed and if they are to go on to the next step in the process. If handbooks or other study materials will be distributed for the written test, include information on where and how to obtain them.

Explain in detail what is included on the physical test and how it will be administered and scored. List the dates and places of test practice sessions, or provide a phone number to call for this information. If child care will be available at practice sessions or at the test itself, publicize that. Provide information on other resources that may be available, either through the fire department or elsewhere, such as through the union, other firefighters' groups, and the community. These resources might include the following:

A videotape about the physical test that explains all of its components and demonstrates techniques that may be used to accomplish them;

Weight-training classes or gym memberships;

Current firefighters who will serve as mentors to candidates;

"Open house" dates or other opportunities for station visits;

Test-taking study sessions, support groups, and/or assistance in finding a weight-training partner.

All recruitment literature, whether it is contracted out to a private company or produced by one firefighter with a desktop publishing program on a home computer, must be neat and professional in appearance. Literature that looks sloppy, contains misspellings and grammatical or typographical errors, and generally appears to have been hastily done and given low priority, makes a negative statement about both your department and your commitment to recruiting women. Current computer and printer technology make it possible to produce attractive, professional materials at very little expense. Take advantage of this. Have someone from outside proofread the material, not only for errors but to see if it will make sense to someone who is unfamiliar with the fire service. If you're using an outside agency, check their text for correct use of terminology. Most non-firefighters don't know that a fire engine is not a fire truck, and even the fire service itself doesn't agree on what a "rescue unit" is.

Posters. Fire departments that have asked applicants how they learned about job openings have often found that posters are a low-percentage effort. (The winners? Newspapers, direct mailing of literature to potential candidates, and knowing someone on the department who told them about the job.) Posters are usually produced outside the department, due to the technical demands of the process; in any case, if you plan to go to the expense of producing and distributing posters, they should be of a professional quality. The typical poster consists of a color or black-and-white photo and a caption such as "Can you fill these shoes?" (with a photo of fire boots), "We're looking for a few good women," or "It takes all kinds to make a fire department."

Posters may be effective recruiting tools if they are well designed to appeal to your target group, and if they are placed in the right locations. They may work better in smaller towns than in big cities where there is more competition for people's visual attention. They may possibly be more effective for long-range recruitment, if they are posted in places where they can remain for some time and more people will see them. As a courtesy to the businesses or other agencies where you are placing the posters, send someone around to collect them once the information is no longer valid. Posters that are placed where they can be easily vandalized or graffitied-over should be checked periodically for replacement or removal.

Videotapes. Videotape technology has created a revolution in information-sharing that the fire service has only begun to exploit. It is relatively inexpensive to produce videotapes and make them available to support your recruitment drive.

One tape should deal with your entry-level physical test, demonstrating each element of the test both separately and as it fits into the testing process. Use women and smaller men among those demonstrating the test items and evolutions, particularly those events that are most affected by height and leverage. Make sure that your test will not change after the videotape is made: the tape must be accurate and give as much helpful information as possible. The tape should be available to loan out. If your department does not wish to handle the paperwork aspects of this, you may be able to negotiate an arrangement with local video rental stores.

You may also wish to develop a videotape for use at orientation sessions and other informational events (booths at Career Fairs, etc.) This tape should talk about the job, and about being a woman firefighter, from the woman's perspective. It might include footage from actual fires, training sessions, "life in the station" scenes, and interviews with women firefighters on the job and at home. Be sure to include a number of women in order to show a diversity of backgrounds, sizes, ages, and personalities.

You may be able to negotiate with local video rental stores to piggyback a copy of your videotape onto specific movies as they are rented. You would provide a number of copies of your tape, to be rubber-banded to the boxes containing the movie tapes. Although a major motion picture about firefighting won't come along every time you want to recruit, this option gives you an opportunity to target very specifically the groups you wish to reach, through the films they are most likely to rent.

Publicity within the community

Following are examples of locations or organizations where you may find it most effective to publicize job openings, place posters and /or leave flyers, and speak to interested groups or to individual women.

Colleges and universities*

High schools, if your minimum hiring age is eighteen

Athletic clubs, teams, and events:
- Universities, colleges, and community colleges
- Women's softball, basketball, and volleyball leagues
- Runs and triathlons
- Area women's powerlifting and body-building competitions
- Self-defense training schools

Gyms and fitness centers

Career fairs

Explorer posts

Military bases and discharge centers

Factories and union offices

Local women's advocacy groups, Women's Commissions, YWCA, 4-H, Girl Scouts, American Association of University Women, etc.

Local minority-employment and tradeswomen's networks

Church and other groups in ethnic communities

Other departments within your city or county government: water, parks and recreation, sanitation, etc.

Area agencies that offer fire recruit training

Local volunteer fire departments, and smaller career departments, particularly if your department does not have a residency requirement

Non-suppression employees of the fire department, and family members and friends of current firefighters and police officers, are two more key sources of potential recruits. These individuals are already familiar with the rewards and demands of the job, and often with the functioning of the department. Use your incumbent personnel as recruiters not only with the public but with people they know personally.

Effective use of the media

The media are the least expensive and most useful resource of your recruitment effort. At very little cost to you, they deliver your message into the homes of thousands of community residents. Not only should you use them for classified ads and public service announcements, but they will often be more than willing to cover your recruitment events as news or feature stories. Smaller-town newspapers are especially useful for this: they usually are in greater need of material, and subscribers often read the paper in great detail.

A number of fire departments have gotten the media to cover their recruitment drives by inviting a woman reporter to go through the physical test, or to spend a day in the station. This can be a bit of a gamble: if the reporter can't complete any of the test events, the job will seem off-limits to women. Providing the reporter with some preliminary training on the events, or suggesting that the station or paper send a reporter who is physically fit and active, may help. It can also be beneficial to have a woman firefighter go through the events (successfully!) at the same time the reporter does, especially if cameras are present.

---

*You may wish to target specific academic areas that you feel will benefit your department: not just fire science programs, but social services or education, for example. Fire departments that provide wildland fire protection might look for candidates with forestry backgrounds.

Encourage reporters to promote the idea that being a firefighter is neither easy nor impossible for women; point to the number of skilled and successful women already on the job in your department and/or across the country. Reporters will usually be grateful for a fact sheet that provides such information about women firefighters, to use as background for their stories. Providing this information will help prevent the media from portraying women firefighters as unusual, even if the women you are recruiting will be the first ones in your department.

Local cable access channels will usually be glad to show your orientation videotape if you provide them with copies. This can also be done in conjunction with an interview and/or call-in show where representatives of the fire department discuss firefighting as a career for women and give information about the current recruitment effort.

The broadcast media are required to dedicate a certain amount of free air time to public service announcements (PSA's). Fire departments can take advantage of these to publicize their recruiting drives. Videotaped television spots should include visuals of women firefighters on the job. Radio announcements, and television spots that do not have visuals, should use women's voices. The text of your PSA's should be consistent with your written handout material; in fact, much of the text can come directly from your literature.

If your department is large, is located in an attractive area, offers particularly good pay and benefits, or is able to hire firefighters from elsewhere on a lateral-entry basis, consider advertising nationally in fire service trade journals or tradeswomen's publications. Keep in mind, though, that monthly periodicals often require a lengthy lead time; be prepared to submit material to them well in advance.

Orientation sessions and open houses

A key element of recruitment involves speaking to groups of potential candidates. Hold orientation sessions at times and locations that are convenient to your target audience. Providing child care during the session will not only make it possible for more women to attend, but will demonstrate your department's commitment to hiring women.

Speakers at the sessions should include firefighters and officers, both women and men, of varying ethnic and personal backgrounds. Their material should include a basic description of what firefighters do, in down-to-earth and unglamorized terms, as well as of the testing and training procedure. Women firefighters should talk about their experience on the job: what it's been like, why they enjoy it. An officer should be present to reaffirm the fire department's commitment to cultural diversity and equal employment opportunity. All speakers should be positive about the job, honest about its demands, and accessible to candidates' questions.

Showing your recruitment videotape at the beginning should reduce the amount of time needed for questions. You also may wish to have personal protective equipment or fire apparatus on hand for candidates to handle or see.

Distribute applications to all those who are interested, if your personnel system allows you to do so. If not, it may allow you to distribute interest cards by which individuals can request that an application be mailed to them. If even this is not possible, at least have each attendee fill out a card with his/her name, address, and phone number. The blank cards should be coded in some method for entry into your data base, so that later on you can evaluate which events were the most productive.

"Open house" sessions provide a chance for candidates to come into selected fire stations, view the apparatus and facilities, and talk with firefighters and officers. When scheduling these events, consider the suitability of the particular crew for such visits, the convenience of the time and location for the target audience, type of apparatus and station responsibilities (SCUBA team, Hazmat unit), etc. Recruitment staff should be present, particularly if you don't plan to take the station out of service. Some fire departments offer open houses on a regularbasis as part of their year-round recruitment effort, as well as for community and neighborhood relations.

Your fire stations should function as recruitment outposts during the entire recruitment drive. Copies of recruitment literature and, if possible, applications, should be available at all stations. All department personnel should be able to answer basic questions about the testing and hiring processes.

Post-application contact

In some firedepartments, the testing procedure alone takes many weeks; months or even years can elapse between the first orientation session and the day a new firefighter is actually hired. Fire departments unnecessarily lose many good candidates during this time. Throughout the application and hiring process, the more contact you can maintain between women candidates and the fire department, or among the women who are trying to get on the job, the less likely you are to lose them and the more likely they are to maintain their interest and motivation. Pre-training programs are one method of maintaining contact. *(See boxes On pages 23 and 25 for examples.)* This time can also be used to help candidates prepare for the upcoming testing process.

Physical test preparation sessions

Physical test practice and preparation sessions provide another method of maintaining contact with applicants. These sessions also accomplish the following:

One fire department's pre-training program

The policy of the Los Angeles City Fire Department since it began hiring women in the mid-1980's has been to provide pre-training to help women who need physical conditioning and strength development in order to pass the department's entry-level physical test. Following are the basic components of their program.

The department's **Tutorial Program** is eight weeks long. Candidates are not screened, and participation is voluntary. The tutorial program allows women candidates to develop their strength and cardiovascular fitness through an intensive program of specificity training with weights, traditional weight training, and aerobic conditioning. The specificity training course was set up under the guidance of kinesiologists and exercise physiologists in 1989, modifying the department's original weight-training program. It involves exercises duplicating the exact physical requirements of the entry-level test.

The program also provides participants with an orientation to the fire department and its operations. Training sessions - one and one-half hours a night, three nights a week - are held at the fire department's training center: the staff consists of one fire captain, two firefighters, and a weight-training coach. In addition, videotaped instruction by women firefighters on the fitness program is provided. Women who can not attend the academy classes have the option of going through the program on an independent basis at private health clubs or school gyms.

Candidates are given diagnostic evaluations to monitor their progress and to determine their ability to perform the exercises that will be required in the next part of the program, preparing them for the physical test. At the end of the program, following a successful strength/conditioning evaluation, medical examination, and personal history examination by the Personnel Department, a group of candidates is selected to continue into the Pre-Trainee Program. Three weeks before the next phase begins, candidates receive a final evaluation, which they must pass in order to proceed. Those who are not selected are permitted to repeat the tutorial program.

The next stage is the **Pre-Trainee Program.** Its participants, all women, receive trainee firefighter salaries (2/3 of beginning firefighter pay). It is an eight-week, forty-hour/week program of "specificity strength/conditioning" training. Four to five hours a day are spent in exercise training; the remainder is devoted to classes and instruction: an introduction to the fire department, oral interview skills, and the safe use of tools. At the end of the program, the candidates take the department's entry-level physical test; those who pass go on to the oral interview, which determines their position on the eligible list. (The requirements of the department's written test are satisfied with successful completion of the academic portion of the Pre-Trainee Program.) Those scoring within an acceptable range - comparable to male candidates being considered for employment-will be considered for one of the next training academies.

A group of ten to twenty men and women then enter the **Trainee Program.** Participants receive trainee salaries. Academic training includes listening skills, readingcomprehension, note-taking skills, and other study habits. It also includes the EMT course and certification. Participants receive a copy of the department's drill manual and recruit training manual and are familiarized with these documents. The physical program includes daily weight training, as well as basic ladder and hose operations. Class size typically has been between eight and 25 trainees; however, due to budget constraints no classes were conducted during the 1991-92 fiscal year. Subsequent staffing shortages led to a modified, accelerated hiring program of much larger classes in 1992-93. The program is staffed by four captains and four firefighters.

The Los Angeles City Fire Department's **Training Academy** is a ten-week basic recruit school teaching fire suppression and rescue activities. Trainees receive entry-level firefighter pay. Staffing consists of four captains and four firefighters.

The Tutorial and Pre-Trainee Programs were created in 1983. The success rate of the programs in their first six years was 15-25%. Following the introduction of specificity training into the program in 1989, the success rate of the two programs increased to an average of 90%. Collectively, as of 1992, the department's pre-training programs had put more than 50 women onto the department.

[Captain James W. Bird is the designer and administrator of the programs described here. See his article in Fire Engineering, *March 1991, for more information.]*

They allow candidates to measure their current fitness levels against what the fire department will require of them, and to identify any areas of weakness;

They provide instruction for candidates in the techniques that can most effectively be used to accomplish the test items;

They give candidates experience in handling the equipment used on the test; and

They give the fire department an opportunity to identify any problems or inconsistencies in the design and administration of the test.

Some fire departments offer both their physical test and practice sessions on a regular basis: for example, practice sessions monthly and the test twice a year. Most departments, however, test only every one or two years, and hold practice sessions during the weeks or months in advance of the test. In either case, the practice sessions should be held far enough in advance that a candidate can go through a practice to find out her/his areas of difficulty, and still have time afterwards to work on those areas before the test is actually given.

In setting up practice sessions, consider the following.

● The equipment, tasks, and sequence of events should be the same as on the actual test, or as close as possible. The convenience of setting up just part of the test for each session may be tempting, but if your test consists of a timed sequence of tasks, participants will not be able to get a realistic idea of the endurance needed to complete the test in the allotted time if they can not run through all of its events.

● Obtain release forms from all participants prior to any hands-on practice or training. You may also require that certain types of clothing or footwear be worn, or you may wish to provide helmets, gloves or other protective items for candidates to wear during practice. If you are providing the equipment, make a full range of sizes available to ensure a safe fit for all participants.

● Firefighters and officers who will serve as staff for the practice sessions should be carefully selected and trained. They should be supportive and encouraging of all candidates, and should offer instruction in all techniques that will be acceptable in actual performance of the test. If candidates attempt to use techniques or methods that are clearly unsafe, the staff should inform them that their method is not acceptable. It is very important to have consistency between the personnel giving instruction at the practice sessions and those who will administer the test, so that candidates are given reliable information.

● Schedule the sessions for different times of day and different days of the week, to increase the number of candidates who can attend. If the testing equipment is relatively portable and easy to set up, consider holding practice sessions at several locations, for the same reason.

Physical test preparation: other resources

Drawing on the resources of a local university, some departments have had an exercise scientist create a task-specific training program for women that is designed around the evolutions on the physical test and the physical requirements of the job and the training academy. (If assessing these three items -the test, the training, and the job - produces three different sets of physical requirements, it may be an indication that re-evaluating your testing and training procedures is in order.) Make sure the program is written down in a way that can easily be understood by someone who is not familiar with weight-training terminology or notation.

Seek community support to supplement what your department can do. Gyms and fitness clubs may be willing to offer membership discounts to candidates who are preparing to take the firefighter exam; make sure the staff members of these gyms receive copies of your training program. Donations from businesses or small grants from community funds may be available to sponsor gym memberships or tuition for a weight-training course at a community college.

Even if a fire department can not, or chooses not to, seek financial support for such efforts, it can do some basic organizational work that will make it more likely that women will maintain their motivation and develop their fitness levels. The department can offer physical conditioning classes itself, or it can provide a sign-up list that will help women candidates find partners to work out with who are motivated by the same goal of becoming a firefighter.

Other forms of preparation

Many fire departments, union locals, and support groups offer or make available "study skills" and test-taking workshops that help candidates prepare for written entry-level tests and for the fire academy. These classes or sessions focus on reading retention, problem-solving, and other skills and tips that assist candidates in the testing process. Local community colleges often have useful resource people in these areas: you may be able to borrow a faculty member to conduct workshops, or find ways to fund candidates' tuition in a study-skills course. This can be of great benefit to women who are re-entering the job market and to those with limited academic backgrounds, particularly if the written test is especially competitive. Workshops on interview skills can also be helpful in preparing candidates for that portion of the hiring process.

Once the testing procedure is over and you have identified the candidates who will enter the next academy, it may be useful to hold a final orientation session, presented by women already on the job to the women

A firefighting "pre-academy"

Captain Jim DeLacey was assigned to run the Oakland (California) Fire Department's pre-academy recruit program in 1991. Although the program had been in operation for several years it was relatively insubstantial at the time. In redesigning it, DeLacey looked also at the department's training academy, where he identified a problem. The firefighters who had come out of the last few recruit classes were good enough firefighters, but they weren't exhibiting the fitness or endurance levels that DeLacey, a former swimming coach, thought they should. Drawing on his background in kinesiology, he developed a program that would promote physical fitness in both the academy and the pre-academy.

The pre-academy is open to women who have passed all of the entry-level firefighter tests and are waiting to be hired. Its goal is to enhance their ability to complete recruit school successfully. Attendance at the sessions is not mandatory, but each student is asked to commit on a weekly basis to the number of sessions they will attend the following week.

The 1991 pre-academy was run on weekday evenings for two hours a night, five nights a week, for six weeks; an average of four to seven women attended each session. The basic fitness program consisted of circuit training, free weights, and running. At the end of the first week, candidates' strength in several muscle groups was assessed by way of maximum repetitions of bench presses, curls, pull-ups, etc. These assessments were repeated at two-week intervals. DeLacey found strength improvements greatest in the middle part of the course: approximately 20% after the second two weeks, and around 25% overall. The greatest improvement was in candidates' running and aerobic capacity: by the end of the pre-academy, participants were running up to 5-1/2 miles three times a week.

The pre-academy also included some basic fireground work. During the first two weeks, participants carried hose packs around the drill ground, learned to raise and extend a 35-foot wooden ladder, developed simulated chopping skills (focusing on body mechanics to produce an effective result) using an eight-pound sledge hammer, hoisted and lowered rolls of 1-1/2" and 3" hose via rope to a third-floor window, and learned safe and effective techniques for lifting large, bulky objects. After two to three weeks, more fireground tasks were added, particularly carrying, "throwing" and raising ground ladders. During this pre-academy, DeLacey found that many of the women were having a problem with the one-person throw of a twenty-foot ladder. His own technique suggestions did not solve the problem. The woman firefighter who was assisting him with the course, who was also a bodybuilder, was able to demonstrate a technique that the women could use effectively.

Morale was high in the pre-academy, which kept the women committed and developed strong ties among the group members. DeLacey's assistant kept in touch with those candidates who were not attending sessions, to maintain contact with them on a regular basis while they were waiting for the regular academy to begin. The entire cost of the pre-academy program to the Oakland Fire Department was one captain's and one firefighter's overtime for the evening hours for six weeks. A multi-station weight machine has now been purchased that will be used in future programs; this particular pre-academy used only the dumbbells and free weights that were available at the time.

DeLacey feels that programs of this type should be run by women if at all possible, and that women firefighters or officers must be involved in some capacity. The trainees who participated in the program, however, said that the complementarity of the captain's and firefighter's personalities, and the difference in their approaches to fitness, were very valuable. Other suggested improvements included having access to first-due equipment so participants can practice on items actually used in the field, instead of older equipment scrounged up from storage. Being able to practice pulling starter cords on generators and saws, for example, would have been helpful. The women suggested that offering sessions at different times of day, and providing access to child care, might have made the pre-academy a possibility for more of the women on the list.

The women who completed the 1991 pre-academy were among nine women who entered the Oakland Fire Department's 16-week recruit training program later that year, all of whom graduated on April 3, 1992, bringing the total number of women in the department to 34, or seven percent of the department's line personnel.

[Thanks to Captain DeLacey and other members of the Oakland Fire Department, particularly its academy personnel, for making this information available.]

who are about to enter training. If your program has been effective up to this point, much of the material may be redundant, but this session will provide one more opportunity to let the new candidates know "what the job is going to be about." It should include information on the role of the recruit, the paramilitary structure of the organization, the function of the chain of command, and the importance of teamwork. It should also be a time when the women can speak freely with each other about the realities and stresses of the job, share solutions and coping skills, and in general offer an introduction to firehouse life. As most fire departments do not employ large numbers of women, new woman recruits are not likely to work with other women once they are assigned to a station. An orientation such as this, beyond providing last-minute information and survival tips, can help create a support network that will compensate for the lack of female role models in the station.

Another means of providing direct support between incumbent personnel and firefighter candidates or recruits is a mentoring program. Individual firefighters or officers, through the department or the union, may volunteer to serve as mentors to candidates, trainees, or probationary firefighters. Mentors offer support, a point of contact within the department, technique tips, and additional needed information. A list of names should be offered to the candidates or recruits so that they can choose the firefighter they personally are most comfortable contacting.

Program evaluation

At the conclusion of any recruitment effort, you should evaluate the effectiveness of the program. The applicants themselves can provide direct and constructive feedback. A space on the application form asking how the person heard about the job, and an evaluation form that attendees at orientation sessions or viewers of videotapes will fill out, can help you determine where your efforts can be improved.

Save the data base you have developed on potential candidates and use it as part of your mailing list for the next recruiting drive. If your eligibility list is likely to last for some time, implement a mechanism that lets you maintain contact with women who are on the list but haven't yet been hired. This lets them know they haven't been forgotten, and increases the chances that they stay interested in the job.

Each member of the recruitment team, and other individuals who have had a significant part in the effort, should submit a written report; this should be followed by a meeting of all who were involved. A final report with recommendations for change and an action plan for implementation of the changes should result and be forwarded to the fire chief or appropriate member of top management. Follow-up letters of thanks from the recruitment coordinator or fire chief to all community members who donated services or money to the recruitment effort are not only polite but may make future donations more likely.

When Judith Livers applied to become a firefighter in 1974, the Arlington (Virginia) Fire Department required her to pass a physical test even though they had never before tested applicants on their physical fitness for the job. Nationwide, as more women began to enter career firefighting positions, the fire service turned its attention to the physical demands of the job and the measures used to test candidates' fitness. As a group, women were often considered incapable of performing the physical tasks of firefighting. For example, a survey of California fire chiefs in 1976 showed that 61% believed women were not physically capable of being firefighters. (Thirty-one percent also said they would not hire a woman even if she were physically and mentally able to do the job.)' The possibility of being faced with incompetent women applicants led many departments to toughen their entry-level standards or to develop physical tests for the first time. Up until a series of court challenges beginning in the late 1970's, such tests were often haphazardly designed and administered by untrained individuals who had no expertise in testing.

The issue of firefighter physical testing is a complex, thorny and often controversial one that raises questions that have no simple answers. There is no perfect solution, no one "best" physical test for firefighter applicants. Within the limitations of fire department budgets, it is very likely that no physical test could predict job performance with pinpoint accuracy.

As a further complication, the validity of any test from one jurisdiction to the next is not guaranteed, since each employer using a test that has an unequal impact on a particular applicant group must validate that test. (The EEOC's Uniform Guidelines on Employee Selection Procedures do permit a validated test from one jurisdiction to be considered valid elsewhere, but only if the user conducts a transportability study demonstrating that the major work behaviors are fundamentally similar for the job in the two jurisdictions.)[2] It is important to note that "validity" has a specific legal definition in this context, and does not simply refer to a general sense that the test is related to the job or contains elements that look like things firefighters do. Designing and validating entry-level tests is the job of exercise physiologists and other experts. The purpose of this section of the manual is to help clarify some of the legal, practical, and ethical issues affecting physical test design and administration, and to provide guidance to fire service managers in assessing the value and impact of an entry-level physical test.

The legal background

The federal law that regulates employment tests, under Title VII of the Civil Rights Act of 1964,3 is defined in the Uniform Guidelines on Employee Selection Procedures issued by the Equal Employment Opportunity Commission (EEOC)." This law, and later court decisions interpreting it, provides that an employment selection device such as a physical test can be challenged if it has an "adverse" or "disparate" impact on a protected group: in other words, if the test disqualifies women or racial minorities in disproportionate numbers.* According to the EEOC Guidelines, a selection rate for a protected group that is less than 80% of the rate for the group with the highest rate will generally be regarded as evidence of a test's disparate impact. For example, if 200 men and 80 women take a test, and 150 men and 40 women pass it, then the pass rate for women (50%) is only 67% of that for men (75%) and could indicate disparate impact.

> The issue of firefighter physical testing is a complex, thorny and often controversial one that raises questions that have no simple answers.

---

*Criteria for protection under Title VII include race. color, religion, sex and national origin. People with disabilities are protected under the Americans with Disabilities Act.

Although "adverse" and "disparate" impact are not explicitly defined in Title VII or any subsequent legislation, the Supreme Court in 1989 tightened the statistical standards for disparate-impact cases.5 The plaintiffs in the case had argued that disparate impact should be determined by comparing the total number of minorities in the geographical labor market with the number in skilled jobs at the employer's cannery. The Court decided that the proper comparison was between the racial composition of *qualified* persons in the labor market and the persons holding the jobs at issue. In a job such as firefighting, where the entry-level educational requirements are relatively low, "qualified" might include everyone with a high-school education. On the other hand, it could be more narrowly defined by a court's ideas about what proportion of women in the population would be physically capable of doing the job. This determination could significantly affect the opinion of the EEOC or a court about how acceptable the percentage of women passing an entry-level test is.

If those challenging a test can show that it has a disparate impact, it then becomes the employer's responsibility to conduct a study that will validate the test. This means that the employer must show that the test accurately predicts which applicants will perform the job better.6 The person or group challenging the test may also show that other tests which would also serve the employer's hiring interests would have less of an adverse impact on the protected group. There are three types of validity, or ways of showing that a test is legal: criterion, content, and construct. *Criterion* validity is proven when the test accurately predictsor is significantly correlated with the actual work proficiency of employees. Content validity is proven when the content of the test duplicates or is representative of actual job duties. *Construct* validity identifies the mental and psychological traits required for successful job performance.7

In defending challenges to entry-level physical tests, fire departments rely on criterion and content validity. Construct validity has been regarded as inapplicable by definition to job duties requiring physical abilities. Validity studies are usually conducted by industrial psychologists or other experts in the field. The actual validation of a test, based on the study, is done in court or by a fact-finder.

Types of tests

Approaches to entry-level physical screening by fire departments fall into four general categories. These are:

> Proxy tests
> Job-simulation tests
> General fitness assessments
> No entry-level physical test

Some fire departments' tests combine elements of the first three types of screening. All four options have advantages and disadvantages in terms of their effectiveness and their impact on women candidates. These will be discussed below.

Proxy tests. Many early challenges to physical testing involved tests that used supposedly "pure" measurements that were alleged to have criterion validity. Tests of this type rely on distance runs to measure stamina, devices to measure grip strength, balance beams to measure equilibrium, eight-foot walls to measure the ability to scale obstacles, etc. The advantages to proxy tests are that they minimize "trainable" tasks (tasks that require practice in order to gain expertise) and tend to give less of an advantage to candidates who have prior experience working with firefighting equipment.

When these "substitute" or "proxy" tests have been challenged, however, fire departments often have not been able to show that the tasks actually measure what they are intended to measure, or have not been able to justify the scoring systems used to evaluate the tasks. For instance, some tests used a dynamometer to measure grip strength. Because the device did not adjust for hand size, it did not accurately measure the grip strength of applicants with smaller hands. When the scoring system for a distance run was examined, in some cases employers could not explain how they determined what would be an acceptable passing score except by caprice or guess." As a result, these tests designed to measure abilities required by the job have been viewed with skepticism by many fact-finders and courts.*

Job simulation tests. Many fire departments use tests that require candidates to perform simulated fireground tasks such as carrying tools and equipment, chopping, simulated forcible entry or pulling ceilings, or advancing hose lines while wearing protective gear. Defending these job simulation tests as having content validity, employers have argued that the test evolutions closely replicate the tasks firefighters actually perform on the job. Fire departments have found the public, unions, courts, and even candidates themselves less willing to dispute a test that, on its surface, resembles what people generally believe firefighters are required to do.

Job simulation tests, like proxy tests, also have their shortcomings. They often include skills that a trained person will perform better than an untrained one, or tasks that require practice in order to gain expertise. This makes an objective assessment of the candidate's raw ability almost impossible. (The EEOCGuidelines caution against the use of trainable items on entry-level selection devices.)'" These tests also often use scoring systems that require the candidate to perform the test at an "all-out" pace rather than a pace actually used on the fireground. Such scoring systems identify the fastest candidate, but

---

*The Americans with Disabilities Act. however, provides for simulated (or actual) task demonstration in order to establish ability within the context of that Act."

Public relations and changing the entry-level test

Most fire departments review their entry-level test periodically to make it more job-related, safer for the candidates, and/or less expensive to administer. If you change your test design around the same time that you announce your intent to recruit women firefighters, your current firefighters and the public may become suspicious of the most recent changes in the test. They may decide that the test is being "watered down" to "let" women on the department. Ironically, this is especially likely when a lawsuit or other challenge has pointed out the deficiencies of the old test.

Clear communication within the department is the first step in dealing with this problem. If your workforce is educated about the changes and accepting of them, the public will usually follow. An aggressive, positive publicity campaign in the media can also help prevent the recruitment and testing process from being poisoned by rumor and misconception. In departments that are unionized, a media campaign will be most effective if it is a joint effort between fire department management and the firefighters' union. Unless the situation has already become highly polarized, a joint task force or other coalition should make a cooperative effort possible.

The public in general has little knowledge of the actual physical demands of firefighting. Fed by media images, people often believe that the harder the testing process is, the better for everyone. Letters to the editor about fire department physical tests usually reflect little awareness of the necessity for valid, job-relevant tests, and represent a general perception that We don't want a 'fair' test; we want the best firefighters." Some examples:

> The fire department's efforts to keep its standards high should be applauded by all, not belittled... by someone who had to file a lawsuit to get standards lowered to get a job.

> Only the most able-bodied should be allowed to become firefighters. (It used to be) that if one could not pass the test, one had to seek less strenuous employment.

In educating your community about entry-level tests, consider the following approaches:

- Point out that the test has changed almost every time it's been given (if this is the case).
- Emphasize any aspects of the new test that measure abilities that the old one didn't, such as aerobic or flexibility components. Offer statistics about firefighter injuries and heart ailments to document the need for such testing. Stress that the new test is therefore not necessarily easier; in fact, it is more comprehensive than the old one.
- Explain the reasons for removing any specific items from the test: for example, the event was removed because it's not done on the job, or not done by one person, or the ability to perform that task is better tested by a new item.
- Note that there are several thousand women working successfully as firefighters and officers on departments of all sizes throughout the U.S. and Canada.
- Emphasize management's commitment to hiring qualified, competent firefighters, and to giving a test that doesn't exclude anyone who can do the job. Avoid saying that you're only complying with federal law in revising your test: you risk sounding like you wouldn't make the changes if you didn't have to.

this determination is meaningless in the absence of data to demonstrate that faster candidates actually make better firefighters. (Scoring systems are discussed later.) Lastly, the events on a job-simulation test may not actually measure what they are intended to measure.

General fitness assessments. Some fire departments, instead of testing for the ability to perform simulated job tasks or proxy events, have adopted tests that assess the overall physical fitness of the individual applicant. The philosophy behind this type of test was expressed by one manager as, "We think our firefighters should have a level of fitness that falls in the top forty percent of the population." (The 40% figure was set by that particular department.) The assessment may include such events as a single-repetition bench press with a maximum

weight, a sit-and-reach flexibility check, push-ups, sit-ups, and dumbbell curls. Scoring for some events may be scaled to the candidate's age, body weight, and gender, to correlate with general fitness standards for each group. This would mean that a 23-year-old male candidate weighing 210 pounds might be required to bench press 140 pounds, while a 34-year-old woman weighing 150 pounds would be required to do only 60 pounds. On the other hand, in flexibility measures, women may be required to out-perform men in order to receive a passing score.

General fitness assessments are not widely used as entry-level tests; where they are, in most cases, elements from proxy or job-simulation tests are also incorporated into the testing process. The primary advantages to this type of test are that it provides a comprehensive view of the candidate's overall physical fitness, and that it does not rely on trainable tasks for its measurements. These tests often will not have a disparate impact on women. Where this is the case, Title VII does not require the employer to demonstrate the test's validity or relevance to the job. (A test must be demonstrably job-related, however, if it tends to screen out individuals or groups protected under the Americans with Disabilities Act[11] or other civil rights statutes.)

Criticisms of general fitness assessments as entry-level tests usually focus on the scaling of pass/fail points to the candidate's age, gender, and weight. While this may be the accurate way to determine how a candidate's fitness level compares to that of others in his/ her category, it runs counter to the "performance" approach that measures the candidate by the demands of the job. Demonstrating the criterion validity of such a test, should that become necessary, would appear to be difficult. Even if the scores are not scaled, general fitness assessments are still subject to the same criticisms as proxy tests in terms of the difficulty of linking pass/fail points to actual job demands.

No entry-level physical test. Some smaller fire departments have never used physical performance tests as part of their entry-level process, relying instead on training and the probationary period to weed out those who can not perform the job. At least one large department in the early 1990's abandoned its entry-level physical test in favor of the screening provided by the training process.

Having no physical component (other than a medical examination) to a firefighter hiring process seems a radical notion. After all, the job requires a certain amount of strength and fitness, even if the exact levels of each are difficult to determine. The idea, however, has its advantages. It does not disqualify applicants on the basis of a test that may or may not be valid. It permits recruits to be trained before their abilities and performance are assessed. And it eliminates much of the expense involved in test development and validation.

It also offers several disadvantages. It runs the risk of putting candidates into firefighter training who do not have the basic fitnessnecessary to go through the training program. This places a burden on the training staff, and is unfair to the candidate who may have quit another job in the sincere belief that s/he had a good chance to become a firefighter. Moving to a "no-test" option may also generate controversy and disharmony within the department.

Having no entry-level physical test does not allow a fire department to escape the responsibility of sorting out qualified from unqualified job applicants. The responsibility is simply merged into the training process, This may only create new problems. It makes the importance of excellent training and highly qualified

instructors all the greater. It also increases the need for a stringent medical physical (within the limits of Americans with Disabilities Act), particularly one that assesses the cardiovascular system and the potential for back injury. An option that can make this approach more workable is to offer candidates a voluntary fitness or physical performance assessment, informing them of the level of performance they will have to reach by the end of training.

One major fire department has adopted a system that modifies this approach. It uses a job-simulation test with a time limit for job applicants that is several minutes longer than the department's performance standard. By the end of recruit training, and annually for the rest of their careers, firefightersmust perform up to the standard.

Developing fair and valid standards

In developing an entry-level physical test, being able to demonstrate its criterion or content validity is only part of the goal. Simply complying with federal law, even if it were always clearly defined, would not necessarily produce a fair or usable test. The EEOC Guidelines, and the industrial-psychologist testing methodology on which they are based, offer a narrow view of discrimination that can not eliminate all the harms caused by past inequities in employment practices.

Why do the Guidelines fail toeliminate bias in testing? After all, they require test development to involve a scientific element, so that a test can not be based on people's subjective opinions about what the job requires. This is a good concept, but what often happens in actual practice is that the "scientific" approach gets distorted or is inadequate. This can happen in many ways.

Failures of the survey process. In order to determine your department's entry-level physical performance standard for firefighters, you must assess what it is that firefighters actually do and how they do it. This is called a job task analysis. It should provide a picture of the job of firefighter, and since job tasks are not always obvious, the analysis must break duties down into their simplest components.* The analysis has an objective component consisting of a list of tasks and how often they are performed, as well as a subjective view of how critical each task is. The test developer gathers information about the job from people on the job ("incumbents"). The more accurate this information is, the better the job analysis will be.

---

*The job analysis provides the basis for the validation study. The analysis must be "a thorough survey of the relative importance of the various skills involved in the job in question and the degree of competency required in regard to each skill. It is conducted by interviewing workers, supervisors and administrators; consulting training manuals; and closely observing... actual performance."[12]

However, the surveys used to determine the critical skills required by the job may not work the way they were designed to. People already on the job, and the evaluators themselves, bring their own biases to the evaluation process; the surveys generally are not designed to eliminate the inaccuracies that these biases produce. It is natural for a dominant, incumbent group to rate the traits they themselves hold in relative abundance as critical to the job, and to underrate the importance of traits they do not possess. Similarly, surveys written by evaluators who have preconceptions about the job will often reflect those preconceptions, in subtle or obvious ways, affecting the resulting data. What should be scientific neutrality ends up being exactly the kind of intuition and traditional belief that the Guidelines intended to avoid, because the values are being determined by the dominant group and not established impartially.[13]

Inaccurate determination of a test's value in predicting job performance. Sex discrimination on the job can make an entry-level test that discriminates against women nonetheless appear to have criterion validity. If women score lower than men on a screening test and then suffer on the job from sexual harassment, denial of training opportunities, and other behavior that diminishes their productivity or has a negative impact on their performance evaluations, a validation study will seem to show that poor performance by women on the entry-level test "predicted" their poor performance on the job.* Unequal treatment can thus perpetuate itself and become a self-fulfilling prophecy, as well as giving men an advantage on the entry-level test.

Confusing common abilities with needed ones. The fact that incumbents pass or do well on an entry-level test does not necessarily mean that the test measures abilities the job requires. A test that measures common traits (ones that many people on the job hold) instead of needed skills (ones that the job demands) may appear on its surface to be valid. For example, if an old, flawed selection device had selected a psychotic workforce, a

---

*In 1988, the Center for Social Policy and Practice in the Workplace at Columbia University issued a report on gender integration in the New York City Fire Department. Columbia used a job survey to measure attitudes of male firefighters and officers towards women firefighters. (All of the women fire fighters had been on the department for five years at the time of the survey.) Men of all levels and experience, regardless of assignment rated FDNY women firefighters as less competent than brand-new male recruits on every firefighting task except skill in community relations. For tasks requiring strength - carrying hose to the fifth floor, making a rescue - men saw the gap in "ability" between new recruits and women as being even wider in favor of new (male) recruits. Further, women's interpersonal relations were rated as significantly less competent than new recruits if the ability to take hazing and pranks was the criterion. Columbia took some small solace in the fact that men who had actually worked with women firefighters rated them slightly higher across all criteria than did men who had not worked with women.[14]

---

profile of incumbents could be used to validate a new selection device that would result in the selection of future generations of psychotic firefighters. Fire department tests may be testing for irrelevant traits: traits possessed by the majority of firefighters because the firefighters happen to be male, not because the traits are critical to task performance .

Institutionalizing traditional values. It is even possible that when we think we are testing for desirable traits, we are actually selecting for harmful ones. At least one fire service manager has openly questioned whether there might in fact be a direct correlation between high scores on the entry-level test and later incidence of job-related injuries.15 A fresh look at traditionally valued behaviors and abilities will produce hiring processes that not only are fairer to applicants but healthier for the fire service as a whole. This applies to attitudes as well as behavior. One model program for firefighter stress management has suggested that:

> Basic fire service attitudes, beliefs and values that are largely masculine and are believed to promote stress, such as omnipotence, a rigid frontier mentality based on strength and toughness, insensitivity, callousness, and extreme risk-taking, may change in a more positive direction with the introduction of women into the fire service.[16]

Designing the entry-level physical test

The ideal in screening firefighter applicants for their physical suitability for the job would be a test that would separate those who, with training, could become skilled and competent firefighters from those who could not. The test would measure each applicant against a standard in an objective way, would not give unfair or irrelevant advantages to anyone, and would not disadvantage anyone for reasons unrelated to job performance. It

Test administration

The best-designed test will not accomplish its intended purpose if it is poorly administered. Your test should be established in full detail before the application period is announced. Your recruitment literature should offer a complete description of what will be on the test, how it will be scored, and how the final eligibility list will be established. Some departments have resisted giving out information about scoring systems, cut-off times, or even about the composition of the test itself; this has a disparate impact on those who may require more preparation time (primarily women). The willingness to spend time and effort preparing for a test is a strong indicator of commitment: the applicant didn't just wander in off the street and decide to be a firefighter. The fact that an applicant prepared for the test should be viewed positively, not as a strike against him/her.

Many fire departments have improved on the practice of requiring candidates to wear firefighting gear during the entry-level test. If candidates wear protective gear, it must be available in sizes that will fit everyone. Applicants who are otherwise qualified may be eliminated by having to use gloves that are two sizes too large, or a poorly fitting SCBA that shifts and throws them off balance; these factors primarily affect women. A better option is to use a weighted vest that simulates the weight of the protective gear and SCBA. These are adjustable to a wide range of sizes and do not give an advantage to one group over another, as the use of firefighting gear can easily do.

Carefully assess the equipment being used on the test events to be sure that it functions as well as equipment used in the field. For example, one fire department's test included a ladder extension. The halyard of the rope on the test's ladder was a smaller diameter than that on ladders used in the field, making it much more difficult to grip effectively. You may choose to make a task less difficult than in the field, compensating for the fact that the recruit will receive training and practice that will improve his/her abilities, but it is hard to justify a task on the test being more difficult than it would actually be on the job.

As discussed in the section on recruitment and pre-training, all techniques that will be permitted on the test should be demonstrated. These should be the same as those that have been taught at any test practice sessions. For the sake of consistency, there should be an overlap of administrative and safety personnel between the practice sessions and the test itself.

would also be economical and relatively easy to administer. Even if no test can precisely meet this ideal, test developers should approach it as closely as possible. Fire administrators should look critically at their department's physical test to see if it truly measures what it is supposed to measure, in the fairest way possible. Three areas where tests often fall short are:

Comprehensiveness;

Whether the test contains trainable elements; and

Whether the test reflects actual fireground practice

Comprehensiveness

A comprehensive test assesses the entire range of physical abilities needed on the job. Many entry-level firefighter tests primarily assess speed (anaerobic capacity) and upper-body strength. *[See Appendix 2 for one example.]* Yet a firefighter's job requires other equally important attributes. Heart attacks consistently cause nearly 50% of firefighter line-of-duty deaths, yet many fire departments have no endurance component to their entry-level tests.[17] Muscular strains and sprains are by far the leading type of injury for fire suppression personnel, on and off the fireground, yet flexibility is often given minimal consideration or is not tested at all.18 The same is true of attributes such as balance, and small motor movements. If, as often happens, women compare favorably with men on the criteria that are not tested, and

candidates are only tested on those areas where women tend to perform less well than men, the results will inaccurately show women to be less qualified for the job than a more comprehensive test would indicate.

Comprehensive entry-level testing is not only fairer to the candidates; it is a benefit to the fire department. The fire service should take advantage of ongoing advances in medicine and exercise physiology in order to improve and refine its ability to evaluate specific physical qualities. Choosing the wrong people at entry level, such as those prone to heart attacks or disabling back injuries, does a disservice to the worker and to the community served.

Trainable elements

Trainable elements on the test should be minimized. Using trainable tasks in entry-level tests is specifically discouraged by the EEOC Guidelines. If entry-level firefighters will all receive basic firefighter training before going on the line, make sure your test does not inadvertently give preference to candidates who have prior experience. (You may choose to favor those with experience by giving extra points through the interview process or elsewhere, but that should not be the function of the physical test.) The tasks selected should be those that require the least training or knowledge of test equipment to perform.

Consider an example of what can happen if an entry-level test contains a significant number of trainable elements. Taking this hypothetical test for a career-level fire department are two candidates. Candidate X has been a volunteer firefighter for two years, so s/he knows some basic firefighting skills: how to handle an axe, raise a ladder, drag hose, open a hydrant, etc. Candidate Y has no firefighting background but has more of the basic qualities the fire department needs in its entry-level candidates: cardiovascular fitness, muscular strength and endurance, flexibility, eye-hand coordination, etc. Because of his/her volunteer experience, Candidate X may outscore Candidate Y on the test and be the one who is hired.

What happened, and why? The candidate who, in our example, was the less qualified of the two, got the job. It happened because a test that should have been predicting job performance instead measured prior exposure to job skills. If fire departments threw recruit firefighters into the field without any training, one could possibly justify such a test. But a real-world fire department that hired both of these candidates would find that, once trained, Y was the better firefighter. S/he was better qualified in terms of basic attributes, while X has already had the benefit of some firefighter training and so will not improve as much during training.

The concept of trainability is not limited to fireground evolutions, but may refer to aspects of cultural pre-training for which physical tests implicitly reward candidates. Job-simulation tests that use firefighting tools and gear give an advantage to men as a group, because men are more likely than women to have worked with axes, ladders, or heavy/bulky objects (ventilation fans, dummies) before, and may be more familiar with performing tasks in bulky gear or gloves. Holding test practice sessions *(see "Recruitment" section)* can somewhat reduce the impact of trainable elements on a test, but it is even more effective to eliminate these elements wherever possible.

Reflection of actual fireground practice

Once trainable elements are minimized, an entry-level test should reflect the way the job is actually performed. This includes pace, time limits, technique, and the selection and sequence of tasks. A test that requires successful candidates to perform at an all-out pace does not represent the way the job is actually done. While many fire departments are careful to set realistic time limits for their tests and to specify that a walking pace be maintained throughout, others use tests that, by design or unintentionally, require an all-out, anaerobic pace.* (If even a brief rest period breaks ups

*In 1989, New York City administered an intense, all-out test scored on speed-to-completion. One male candidate died during the test, and more than thirty other men were hospitalized or treated for renal failure. Other cities that have administered comparable tests have seen similar medical problems result.

a "long" test designed to measure aerobic capacity, stamina and endurance, it may instead become two or more anaerobic/speed tests.) Although actual fireground performance calls upon the firefighter's aerobic capacity, these tests require speed-to-completion performance and therefore measure only anaerobic capacity. Their pace and intensity do not match the way firefighter tasks are really performed. Women as a group tend to compare favorably with men on aerobic capacity, but may be disadvantaged by tests that over-emphasize anaerobic ability.

Sequential tests that lump all tasks together under an overall time limit do not measure minimal ability in any single task. A candidate could perform one task at a speed that would be unacceptably slow on the job, but still pass the test by completing other tasks at speeds that might be unnecessary on the job.

The diversity of firefighter sizes and shapes, and the corresponding variations in effective fireground techniques, must be considered in designing entry-level tests. Different firefighters perform fireground tasks in different ways. This means that, on the job, the techniques used to accomplish a task will vary depending on who is in the workforce. For example, firefighters who have a lower center of gravity or greater strength in the lower

An alternate hiring process

The Kern County, California, Fire Department developed its Work Experience Firefighter Trainee Program in the mid-1980's in order to increase the number of women and minorities in suppression positions on the department. KCFD in 1984 had 487 line personnel, of which 51 were minorities and none were women. The Work Experience Program was designed to offer a group of firefighter applicants not selected through the department's standard testing process a year of on-the-job training and testing as an alternate entry path. The program has been successful and is a good example of union/management cooperation: both parties were involved in the development of the program and both have a stake in its success.

Kern County's entry-level testing procedure for firefighters normally attracts 1300-1500 applicants. The written test is given first; only the top 50 or so scorers on the written exam are permitted to go on to the physical test. Women and minorities who are not in the top 50 are offered the option of the Work Experience Program. Candidates go through a second written exam, scored pass/fail, and then to an oral board which establishes their ranking on a list for entry into the Program. They also take the entry-level physical and must score within a "trainable window:" high enough that it can reasonably be expected that they will be able to pass the test at some point during their year of training.

The trainees then enter the regular ten-week recruit training course along with recruits hired from the regular list. (By agreement with the union, this is on a one-for-one basis.) After completing the course, trainees are assigned to one of four Program training stations, which are selected for the types of apparatus housed there, a high level of calls, and generally because the captain or engineer on the shift at that station is interested in being involved in training. Trainees work a standard firefighter shift and receive two-thirds of firefighter base pay with no benefits (including sick leave). Once a month, all trainees are brought to the academy to receive additional training. At the end of the training year, trainees must take and pass the department's physical test. If they fail to do so, they are terminated; if they pass it, they go on to become probationary firefighters.

The program has been successful both in terms of numbers and acceptance: in its first four years, nine women and 23 minority men became Kern County firefighters. The firefighters' union supports the program and agrees that it produces skilled and competent graduates: "These people have proven themselves twice over." Representatives of several other firefighters' unions have visited Kern County to observe the program's operation.

The program is also controversial, and not without its administrative problems. The low pay and lack of sick leave for trainees who are performing essentially the same job as firefighters is one point of criticism. Those within the department cite the need for improvement in a few areas: a better method of selecting the officers who staff the training stations (to improve consistency and quality of training), an option for an "early turn-out" if a candidate can pass the physical test before a full year has elapsed, and improved recruitment to field more women and minority applicants.

Most significant, however, is the seeming paradox that the recruits who enter firefighter training from the regular hiring list and the trainees who come in through the Work Experience Program are often indistinguishable during the academy training: that is, the "regular" recruits do not consistently out-perform or outscore the Work Experience trainees. This strongly suggests that the testing process as it is currently administered is not a reliable predictor of job success. Trainees who do better in recruit school than those hired off the standard list must still complete their year at two-thirds pay before they can begin their probation, even if they are able to pass the physical test.

The Work Experience Program nonetheless is an innovative method of increasing the numbers of women and minority men on a fire department, particularly in its display of labor/management cooperation.

[Thanks to the management, union officials, and members of the Kern County Fire Department for making this information available.]

An argument for linkage

Developing a test to measure an applicant's ability to perform any job requires a thorough knowledge of the job in question. Once the job analysis has been completed and the list of important (frequent or critical) tasks is compiled, test developers can, with fire service input, determine which of the tasks to include as part of a physical performance test. This selection is the most difficult and controversial part of test development.

On an entry-level exam, job applicants should not be expected to perform tasks that firefighters do not perform on the job. Firefighter applicants should also not be expected to perform at a higher level than incumbents. Regardless of the tasks selected for an entry-level test, the performance standard must be linked with the incumbent's level of performance. The lowest level of acceptable and effective incumbent performance is the standard for applicant performance.

The assessment of incumbent performance will provide a framework for establishing time limits for completing the tasks that make up the test. An all-out, fast-as-possible pace is unsafe, unnecessary, and does not reflect real firefighting practice. Applicants who can perform necessary tasks at a pace that is necessary to the job and at which incumbents can and do perform, must be considered capable.

Basing entry-level standards on incumbent performance has a number of advantages. It is relatively easy to compile a job task analysis and to generate data on the performance of those on the job. Complex statistical or physiological date need not be collected, which reduces administrative costs. Most importantly, linkage of entry-level standards to on-the-job performance leads to the development of a single performance standard for all firefighters within a department.

---

part of their bodies (primarily women) may find it more effective to use a different technique for pulling a ceiling or cutting a roof than those who have greater strength in the upper part of their bodies (primarily men).

One place this has an impact is in the sequence of events on the test. On the job, women firefighters often use their lower bodies as well as their upper-body strength in forcing entry. But if the sequence of tasks on the test exhausts all lower-body muscles by the time the candidate reaches the forcible-entry simulation, candidates will be required to perform the task in a way they might not actually perform it on the job - that is, by using their arms only. The test will then measure the candidate's ability to perform forcible entry with his/her arms, and not the ability to accomplish the task in the way that is most effective for the individual. Those who have less upper-body muscular development will be unnecessarily disadvantaged.

Another place technique can have an impact on test design is where events on a job -simulation test are based on techniques that not all firefighters use. Fore example, a job-simulation test might require candidates to use the same technique to lift a dumbbell that men use to lift ladders on the job. Since other equally good ways to lift ladders exist that women may use more effectively, the dumbbell lift could eliminate women candidates who could actually perform well on the fireground.

A fire department can defend its entry-level test by showing a positive correlation between a firefighter's performance evaluations on the job and his/her performance on the test.* If no women are currently on the job, what appears to be a positive correlation between job performance and entry-level test performance may actually only correlate a test biased in favor of men with a workforce that is predominantly male. It might also correlate low job performance by women who have not been trained in appropriate techniques with low performance on such an entry-level test.

If a more diverse workforce alters job skill performance by using a variety of techniques for lifting ladders, the correlation between the exam and fireground performance can go down markedly. And as the workplace itself changes - increasing hazardous materials incidents, lighter and more sophisticated tools and equipment, increased emphasis on fire prevention -selection devices must also change to reflect new work tasks and realities.

Scoring systems

The scoring system used to rank candidates is another crucial factor in determining the fairness of a test.19 Entry-level physical tests are most often scored on a pass/fail basis. Other options include rank-ordering and "banding."

Many fire departments have adopted the philosophy of a "qualifying" physical test. Employers scoring their test on a pass/fail basis must fulfill two requirements in justifying the cut-off score. First, the test must be "reliable:" that is, if it were graded, or the person evaluated, by two different people, the results would be fairly similar. Second, the employer must have a "justifiable reason" for adopting the score.[20] Pass/fail physical testing generally has the least disparate impact on women, of the three options under discussion here.

---

*How "positive" or statistically significant that correlation must be has never been defined by Congress, the EEOC. or the courts.

Other employers, adopting the "more is better" philosophy, score their test on a rank-ordered basis. Some group candidates into "bands" based on their scores, and consider all individuals within each band to be equally qualified. This recognizes and compensates somewhat for the impreciseness of the selection device, but it can permit a one-second difference in time to make a five- or ten-point difference in a candidate's score.

If a test is used to rank applicants, the employer must show that higher scores correlate with better job performance. The Supreme Court has held that where the disparity of impact is greater at high passing scores than low, any alleged correlation between higher scores and better job performance must be "closely scrutinized."[21]

Firefighters of both genders and of all shapes, sizes, and colors have demonstrated successful performance on the job. Studies demonstrate that the trainable component of many exams is a greater disadvantage to women as a group. Given these facts, it makes sense to doubt the reliability of rank-ordered scores. A ten-second difference is often all it takes in a rank-ordered, speed-to-completion test, to separate someone who's hired from someone who isn't. It would be extremely difficult to develop a test so accurate that a ten-second difference in the overall score would really predict who would be a better firefighter.

A question that is often asked is whether one can justifiably require job applicants to perform tasks (or perform at speeds) that firefighters on the job can not themselves perform. The rationale for rank-ordering and for high cut-off scores for pass-fail tests has traditionally been based in part on common assumptions about the effects of age. Fire departments have argued that because workers' physical abilities "naturally" deteriorate with age, employers are justified in requiring greater physical abilities from new hires than the level of physical fitness actually held by incumbent firefighters.

This assumption is becoming increasingly questionable. The EEOC in 1992 was scheduled to issue findings arguing that Congress should revoke the fire service's exemption from the Age Discrimination Act. The study that was done to research the issue concluded that age was not a reasonable predictor of employee performance in the fire service: that "accumulated deficits" in firefighter physical ability were only marginally associated with age.[22] As attitudes change on the issues of diet, exercise, and smoking, it will be increasingly difficult to argue that the fire service can maintain separate standards of physical fitness for new hires and incumbents. Rather, the trend in the fire service will undoubtedly be to have one standard for firefighters, as reflected in the recent revision of NFPA's 1582 "Standard on Medical Requirements for Fire Fighters."

Eliminating bias in testing

Fire department managers should examine entry-level testing processes carefully for elements of bias that give any group an unwarranted advantage. Bias is not always obvious. It may be concealed in "substitute indicators" such as using speed as the primary scoring criterion. It may be hidden in the job task analysis, or in test events that do not measure what they are designed to measure because they allow only one technique in the performance of a task, or incorporate elements on which candidates can improve with training. It can also be as simple as candidates being required to wear gloves on the test, when the gloves do not fit all of the candidates.

Fire departments that use job-simulation tests to assess candidates' performance have found these "look-alike" tests to be widely accepted. On the other hand, some exercise physiologists have developed job standards that are based on complex measures of the energy costs associated with fire suppression tasks, and claim that their criterion-linked standards are more defensible in court. Whatever the type of test used, where there is disparate impact, the test may be challenged in court, and the fire department using the test bears the responsibility of defending it.

Many fire departments have found that establishing the validity of a physical test is a lengthy, difficult and expensive process. The employer must justify all aspects of the test, from the selection of tasks to be performed to the time allowed for completing the tasks. If the test is not scored on a pass-fail basis, the employer must demonstrate that a higher score correlates with job performance. Regardless of who wins in court, the enormous expenses of time, money and morale can produce a no-win situation for both sides in the legal challenge. For the good of the fire service, as well as fairness to applicants, traditional physical performance tests and standards should be re-evaluated.[23]

Notes:

Affirmative Action Committee, California Fire Chiefs Association; "CFCA Survey of Women Fire Fighters," 1976.

[2] *29 C.F.R. §1607.7*

[3] 42 U.S.C. §2000e et seq, as amended.

[4] 29 C.F.R. §1607: Uniform Guidelines on Employee Selection Procedures.

[5] *Wards Cow Packing Co., Inc. v. Atonio, 109 S.* Ct. 2115 (1989). But *see Dothard v. Rawlinson, 433* U.S. 321 (1977) where the Supreme Court explicitly recognized that a plaintiff could establish a statistical showing of disproportionate impact with "potential applicant" data.]

[6] The Uniform Guidelines permit an employer to use a test even though it is not at the moment fully validated if 1) "substantial evidence" of validity is available, and 2) a validation study is in progress that will yield the missing

data in a "reasonable time" (unless a study is not feasible). [29 C.F.R. §1607.5(J)] If the test is subsequently found invalid, adherence to this section will not legally insulate the employer from a charge of discrimination.

[7] These definitions were developed in reliance on standards developed by the American Psychological Association, the American Education Research Association, and the National Council on Measurement in Education in 1974. [29 C.F.R. §1607.5(C).] Published by the EEOC in 1978, the definitions have not been revised or reviewed for possible conflicts with later developments in testing and exercise physiology.

[8] *Berkman v. Koch,* 536 F. Supp. 177 (E.D.N.Y. 1982) *aff'd* 705 F.2d 584 (2d Cir. 1983).

[9] 29 C.F.R. § 1630.

[10] *29 C.F.R. § 1607.5(F).*

[11] *29 C.F.R. §1630.14(a).*

[12] *Vulcan Society v. Civil Service Commission, 360* F. Supp. *1265,* 1274 (S.D.N.Y. 1973).

[13] *See Kilgo v. Bowman Transp., 570* F. Supp. 1509 (N.D. Ga. 1983), *aff'd 789* F. 2d 859 (11 th Cir. 1986). Job analysis was found wanting for several reasons, including the fact that it profiled the trucking industry according to its current overrepresentation of male employees. See also *Vulcan Pioneers v. N.J. Dept. of Civil Service, 832* F. 2d 811 (3d Cir. 1987).

[14] Center for Social Policy and Practice in the Workplace, "Gender Integration in the New York Fire Department: A Review and Recommendations," Columbia University School of Social Work, 1988.

[15] Osby, R., "Guidelines for Effective Fire Service Affirmative Action," *Fire Chief,* September 1991, p. 53.

[16] *Stress Management: Model Program for Maintaining Firefighter Well-Being,* produced by the IAFC Foundation for FEMA/U.S. Fire Administration, 1991, p. 35.

[17] Washburn, Arthur E., Paul R. LeBlanc, and Rita F. Fahy, "Report on l991 Fire Fighter Fatalities," *NFPA Journal,* July-August 1992, pp. 40-54.

[18] Karter, Michael J., Jr., and Paul R. LeBlanc, "U.S. Fire Fighter Injuries - 1990," *NFPA Journal,* November-December 1991, pp. 43-48.

[19] The Uniform Guidelines address cut-off scores at 29 C.F.R. § 1607.5(H).

[20] In *Gillespie v. Wisconsin,* the court said the employer must use a professional estimate of the requisite abilities or at least a logical "breakpoint." 771 F. 2d 1035 (7th Cir. *1985).*

[21] *Guardians Assn. v. Civil Service Commission , 630* F. 2d *79* (2d Cir. 1980). Also see *Pina v. City of East Providence, 492* F. Supp. *1240* (D.R.I. 1980); *Brunet v City of Columbus, 642* F. Supp. 12 14 (S.D. Ohio *1986),appeal dismissed, 826* F. 2d 1062 (6th Cir. *l987), cert denied, 485* U.S. 1034 (1988); *Zamlen v. City of Cleveland, 686* F. Supp. *631* (N.D. Ohio 1988), *aff'd, 906* F. 2d 209 (6th Cir. 1990), *cert denied, 111* S. Ct. 1388 (1991).

[22] Landy, Frank J., "Research on the use of fitness tests for police and firefighting jobs," The Pennsylvania State University Center for Applied Behavioral Sciences, 1992.

[23] This section has part of its philosophical basis in the excellent analysis of testing and EEO by Kellman, M., "Concepts of Discrimination in 'General Ability' Job Testing," l04 *Harvard Law Review* 1158 (April 1991).

The quality of a fire department's recruit training is crucial to every firefighter's chances for success on the job. It can be doubly so for women recruits. Women often enter recruit training with more to learn. They may have less familiarity with the fire service than do many of the other recruits, or less past experience working with ladders, axes, power saws, and other tools and equipment that are commonly part of men's cultural pre-training. This means that they must work and study harder in order to complete the training successfully. It also means that any inadequacy on the part of the curriculum or the instructors will have a greater impact on them.

The quality of a fire department's recruit training program should be assessed for the following characteristics:

A comprehensive curriculum that teaches the skills and knowledge recruits will need on the job;

Presentation of a variety of techniques for accomplishing physical tasks;

Effective teaching methods;

Qualified and motivated instructors; and

Consistency in how material is presented and how recruits are rated.

The training curriculum

The contents of basic firefighter training should be drawn from the tasks a probationary firefighter is expected to be able to perform and the knowledge s/he is expected to have. The curriculum design should be based on a job task analysis.

The fire service has a strong tendency to allow evolutions, skills, or techniques to remain as part of recruit training even when they are no longer relevant to the job. Instructors sometimes rationalize this by describing the requirement as good for "toughening the recruits" or for "building character." The real reasons such items persist may have more to do with tradition, with what the instructors had to do when they were recruits, or with a hidden desire to make the training as difficult as possible. Evolutions that are not required by the job have little or no place in a realistic, productive training program. The time allotted for recruit training is usually limited and can be much better spent on the skills and knowledge the firefighter actually needs. And poor performance on an irrelevant evolution can unjustifiably brand a recruit (to others and, most damagingly, to him/herself) as being "borderline" or "having problems."

There may be some justification for including items that help teach skills or develop strengths that will be needed on the job. Calisthenics and aerobic exercise, for example, are part of many recruit academies. While firefighters do not perform mile runs on the job, running is an effective way of improving the aerobic capacity that is crucial to a firefighter's job performance.

The training program should be comprehensive in including abilities that are needed on the job, and should exclude those that are irrelevant. Eliminate or revise any areas that implicitly require recruits already to possess certain skills or types of knowledge. These hidden requirements may exist not because this "pre-training" is really necessary, but because the type of recruit you've always had has always known these things.

One woman firefighter described her experiences:

When I was first being taught to open a hydrant, they handed me the hydrant wrench and told me, "You use this just like you would any other wrench." I had never, to my absolutely certain knowledge, ever held a wrench of any kind before. They tell me to cut along the floor joists; I have never wielded an axe, never heard the term "joist." We talk about the dangers of vehicle extrication; I am clueless as to the internal workings of a car. Five months into the job, with all new gear that

The training program should be comprehensive in including abilities that are needed on the job, and should exclude those that are irrelevant.

genuinely fits, and strength that is adequate for most situations, I am still perpetually handicapped by my ignorance of information that boys in our culture just seem to absorb.[1]

Find ways to impart such skills or information to recruits who don't possess them. One fire department, for example, hired university personnel to come in and teach study skills to its recruits. Others offer extra sessions on basic tool use or incorporate training on residential heating systems and basic electrical systems into their curriculum.

Look at both the job and your training to identify new areas of need. Fire departments have been undergoing rapid change; your training curriculum has probably changed as a result. Hazardous materials training has undoubtedly been added because of increased fire department awareness, the need for hazmat response, and related legal requirements. But have other elements such as cultural diversity training, anti-harassment education, communications skills, and conflict resolution also been incorporated, to reflect other new workplace needs?

Technique

Assess your current techniques for performing tasks in the same way that you have looked at the training curriculum. To require recruits to perform a task in the most difficult way possible simply *because* it is difficult is a legacy of the boot-camp methodology of firefighter instruction. Some women firefighters have been instructed that they must operate a 2-½" line without being able to rest it on their hip; others have been informed that it's "cheating" to bend one's knees when pulling a halyard to extend the fly on a ground ladder.

As one battalion chief said, "Any training instructor who thinks there's only one way to perform an evolution is missing the point." Just as was done for the entry-level physical test, techniques for smaller people, shorter people, and people with different body mechanics and leverage should be incorporated wherever possible into physical skills and evolutions. For many skills, several different techniques are possible without compromising safety. Demonstrate the different techniques equally; don't give the impression that one is somehow inferior or should only be used if the recruit is having trouble with the "normal" or "regular" method.

A firefighter training class for women

At the request of the Jacksonville Fire Department, the Florida Community College at Jacksonville (FCCJ) in 1990 offered a women-only firefighter certification class, the only such program ever run in Florida. (FCCJ had earlier run a Black Firefighters' Program that placed 75 of 88 participants with JFD.) The women's class was designed to increase the number of women on the Jacksonville department, and it succeeded in doubling that number from eight to sixteen.

Florida is one of a few states that permit people to take firefighter training in a vocational-school setting before applying to a tire department. FCCJ offers the training on a regular basis, but was asked to make this class open to women only. "We feel that an environment with only female students would be less threatening, and students would feel more at ease requesting help if needed," the Jacksonville Director of Public Safety said.

Following a publicity campaign and orientation session, interested women were given written and physical performance tests. Those scoring below the cutoffs on the written tests were allowed to continue with the program while attending remedial classes in English, reading, and/or math. The physical test was given as a diagnostic tool to determine areas where work was needed, and to establish a baseline to document later improvement. The women who passed the test were allowed to go directly into the certification program; 69 of the remaining 79 women entered a sixteen-week physical training course designed by a physical education professor at FCCJ.

Halfway through the course, the students took the physical test again. All of the 39 women remaining in the program improved their times by two minutes or more; nineteen passed the test, and six came within five percent of a passing time. Based on these results, it was decided to begin the firefighter certification class in mid-April. Both day and evening sessions were offered; 33 women attended in the daytime (eight hours a day, five days a week) and six came at night and on Saturdays, for a total of 372 hours of training over the course of the program.

Joe Fowler, Director of the Public Services Program at FCCJ, was impressed by the ability and dedication of the women. "At one point, I received a phone call from the coordinator telling me that I needed to be at a ladder evolution test the next day because it was felt that six or more of the students would fail. I showed up bright and early, and so did the women; not one failed."

A variety of funding sources made it financially possible for women to take the course. Only the six women in the evening classes were working full time during the program, and many of the women had quit their jobs in order to participate. Seventeen women took advantage of a scholarship loan program from FCCJ underwritten by a local bank, that provided a $200/week stipend for the nine weeks of training, to be paid back over three years at $55/month. Several women received book fees and tuition from a non-traditional women's employment program or a local Private Industry Council. Some women also received funding to cover child care expenses.

Twenty-four women successfully completed the course. For the fifteen who dropped out or failed, the primary reason was the written portion of the course. Fowler noted that this was unusual, and speculated that so much emphasis might have been put on the physical conditioning aspects of the course that its mental and intellectual demands might not have received enough attention.

The placement rate with the Jacksonville Fire Department of the women who graduated from the course was not as high as had been hoped, due to a number of problems and delays, and also to a priority shift that led the department to hire certified paramedics before considering other candidates. Nevertheless, at least eight of the women were hired by JFD later that year, and the department was carrying about 60 vacancies, with the hiring list valid for two years. Several other women in the class were hired as firefighters by other departments.

[Thanks to Joe Fowler of the Florida Community College at Jacksonville for making this information available.]

The use of alternate techniques is usually controversial at first. Officers and instructors tend to view the traditional methods as "right," and then find some pretext for justifying them as "safer." Real issues of safety are often hard to separate from what is simply a desire not to make things easier. For example, many women firefighters have found that using a reverse grip on a ladder halyard (with the thumb edge of the hands facing downward) allows a more efficient use of strength. Some departments have accepted this technique; others have expressed concern that the resulting rotating motion of the wrist might lead to injuries.

In reality, techniques that are moreefficient for smaller, lighter people often also end up being safer and more efficient for others. "If the big guys use the same technique as the smaller people, they distribute the work more throughout their body, fatigue more slowly, and have less potential for pulling muscles," the training chief for one department observed. When evolutions require coordination of several people's efforts, some standardization of technique may be necessary for efficient operation, but efficiency, effectiveness and safety should be the guiding criteria, not "how we've always done it."

Delivering the training

Firefighter training has traditionally been modeled on team sports and the military. Reasons for this include:

They are what the instructors themselves know, from their own backgrounds and from their training as firefighters;

They are usually effective with recruits who have those backgrounds;

They are consistent with other paramilitary aspects of firefighting.

Neither of the first two is a compelling reason to continue to train in a particular way, if the people you're training have different backgrounds than those for whom the training method was designed. Most fire academies are now finding it productive, regardless of who the recruits are, to emphasize learning over intimidation. Where the old approach was, "Let'ssee who can survive this course," the new one is, "We're going to do all we can to teach you to be a firefighter." If particular aspects of paramilitary culture are important, they can be taught specifically, such as when to take an order without discussing it, or how the chain of command works, but the training setting should be responsive to the students' needs.

Instructors

Instructors should be selected for their ability to teach and to communicate, their sensitivity to the needs of a variety of students, and their willingness to offer extra assistance to students who request it. They should have a positive and supportive attitude towards training and towards the recruits. Instructors should not be assigned to the training center as punishment, because of a temporary or permanent disability, or in order to get points on their next promotional process.

The instructional staff should include women, minority group members, and people of different heights and sizes. This is particularly crucial for classes that have women and minority recruits, but it is an important general practice as well. Academy training is the recruit's first point of extended contact with the fire department. It is used in many ways to instruct recruits in basic fire department reality -discipline, punctuality, etc. -and it should also give the clear message that the fire department is a culturally diverse workplace. All instructors should be trained on cultural diversity and anti-harassment issues as well.

Consistency and fairness

Consistency in the presentation of information to recruits is very important. This includes both curriculum items and information about how the class will be run. The criteria for passing or failing the academy should be clearly explained to all recruits and strictly adhered to by the instructors and staff. There should be no sources of "inside information" regarding tests, evaluations, or techniques; all recruits should have equal access to any useful knowledge.

Ensure that your instructors permit recruits a chance to receive full instruction and practice on all tasks before they are evaluated, and that each student completes each task for which they are responsible. Time constraints sometimes lure instructors into "checking off" students who in fact have only watched someone else perform a task; this is unfair to the student and to the department. If tests or evaluations will be given without notice, students must be clearly warned that this is possible, and should be informed as to what material is subject to such evaluation. Documentation on the performance of all students in both written and practical skills must be comprehensive and consistent. Some departments have begun using videotapes of recruits' physical performance both as learning tools- the instructor can sit down with the student later and point out any problem areas-and as part of the evaluation process.

Notes:

[1] Brown, Cathy, "One Crazy Person Running In," *WFS*

The ability of a fire department to retain women firefighters will depend, in large part, on whether it develops and implements clear and fair policies in areas that have a unique or particular impact on women. These areas, to be discussed on the following pages, include reproductive health and safety, child care, firefighter marriages, hair length and grooming, and fire station facilities. The very important topic of sexual harassment will be discussed at length in the following section.

It is important to have these policies in place before the first women are ever hired. There are four reasons for this:

- Policy development and implementation can be handled carefully, not rushed into place at the last minute.

- The department shows that it is prepared for women to be firefighters and supportive of their presence. Single-gender firehouse facilities, male-based hair standards, and/or a lack of a maternity policy, all give the clear message that women aren't really expected to make careers within the department.

- Any controversy surrounding the implementation of a policy will be focused on management and on the policy itself; if women are already on the job, they often become scapegoats and are blamed for unpopular policies. Additionally, if policies are inadequate, it is often the women who must seek to have them changed, thus drawing unwanted attention to themselves and again being blamed for the change.

- All members of the organization know what they can expect. The "we'll deal with it when it happens" approach to pregnancy, for example, means that a woman who is being responsible about family planning will not be able to start her family, since she has no idea what will happen to her if she becomes pregnant. Women whose pregnancies are unplanned will be subjected to unnecessary stress and the possibility of financial hardship. Similarly, firefighter couples who wish to marry are entitled to know what the impact of their marriage would be on their work assignments and promotional opportunities.

In paid fire departments, some of these policies, such as maternity leave, may be subject to collective bargaining with the firefighters' union and thus can not be unilaterally implemented. In such situations, it is important for the union's negotiating team to be informed about the issues and energetic about pursuing them. If written policies have not been negotiated or implemented, management should clarify what the current policy is, or what the lack of policy means. For example, what will happen to a firefighter or recruit who becomes pregnant? Similarly, all other fire department policies should be reviewed for their impact on women. For example, if all fire departments in the area have residency requirements, what would happen if two firefighters from different departments got married?

Many volunteer fire departments do not have written policies in these or any other areas. There are compelling reasons for this to change, particularly if volunteers receive compensation for their time and can be considered employees under state or federal law. For all fire departments, having comprehensive, impartial policies in place lets all personnel know what the ground rules are and makes it more likely that each individual will be treated fairly. This enhances recruitment, morale, retention, performance, and almost every other aspect of fire service life.

Comprehensive, impartial policies... let all personnel know what the ground rules are, and make it more likely that each individual will be treated fairly.

Up until the late 1970's, the fire service gave little thought to maternity or family issues. Although women had been volunteer firefighters for many years, most fire chiefs, especially chiefs of paid departments, believed that they would never have to deal with women on the job. Many departments that did have women firefighters dealt reactively rather than proactively with new issues and situations as they arose. Instead of anticipating the future and planning for it, fire chiefs were often caught off guard by change, with results that served no one's best interests.

This was especially true regarding the issue of maternity . Until fairly recently, many employers in all types of work refused to deal with maternity as an issue in the workplace. Only a generation ago, motherhood and work outside the home were considered mutually exclusive. If a woman had a job at all, it was expected that she would quit once she started having children. Employers were allowed to treat pregnant employees in any way they chose, and the law offered no protection for these women. As recently as 1964, forty percent of all employers terminated workers who became pregnant.

Times have changed. Two-income families are now the rule rather than the exception among families with two parents, and most women who work do so because their income is needed to support themselves and their families. More than half of all women with a child under one year old are working outside the home. Fewer than ten percent of all U.S. families now reflect the traditional model of a husband who works and a wife who stays home to raise the children.[1] Yet policies in the workplace that support employees' dual roles as breadwinners and caregivers to children (and, increasingly, to aging parents) have been slow to develop.

The development of workable policies in the fire service has been more challenging due to the nature of the job: the odd hours, the hazardous work environment, the near impossibility of taking years off from the job for the purpose of child rearing. Because the issue is inherently complex, and because at first only a few women, widely scattered around the country, were affected, the overwhelming tendency of fire departments was to avoid the issue as long as possible, and often to find ways to encourage (or coerce) women who did become pregnant to quit their jobs.

A 1986 survey of seventy fire departments nationwide found that only ten percent had a maternity or pregnancy policy specifically designed for fire department employees[2] Forty-four percent had no maternity provisions whatsoever. Four years later, the picture had changed only slightly. A 1990 survey showed that between a quarter and a third of fire departments in the country that employed women had no written policy regarding maternity, and an additional ten percent had only a city-or county-wide policy that did not specifically address women firefighters. Between thirty and fifty percent of departments reported having some sort of written maternity policy for women firefighters, although in many cases this simply consisted of language specifying that pregnant women could use sick leave and vacation time, or that departmental off-duty injury leave provisions applied to pregnant women firefighters.[3]

This section on reproductive issues will discuss legal aspects of the maternity issue, clarify the difference between maternity policies and pregnancy policies, and explain the concept of gender-neutral parental leave.

Pregnancy and the law

One federal statute and two court decisions have set the course for the current discussion of maternity policies in the workplace. The law is the Pregnancy Discrimination Act of 1978 (PDA), an amendment to Title VII of the Civil Rights Act of 1964. The PDA broadens the definition of sex discrimination set forth under Title VII to include discrimination based on pregnancy and childbirth. It states:

> Women affected by pregnancy, childbirth or related medical conditions shall be treated the same for all employment related purposes, including the receipt of benefits under fringe benefit programs, as other persons not so affected but similar in their ability or inability to work...[4]

This law applies to all employers with fifteen or more employees, and to employment agencies and labor organizations.

The PDA was the first significant piece of legislation to deal with maternity in theworkplace. Its main purpose was to prevent arbitrary and discriminatory treatment of pregnant employees. The PDA guaranteed that pregnant women would have access to the benefits already in place for other workers. For example, company health insurance plans could no longer exclude coverage for pregnancy and childbirth. An employer could not refuse to hire or promote a woman solely on the basis of pregnancy. Disability caused by pregnancy had to be covered under an existing disability program. If the employer gave temporarily disabled workers light duty, it was also required to give pregnant employees light duty. Most importantly, a woman could not be arbitrarily fired from her job if she became pregnant, nor could she be required to take an extended leave that was not deemed medically necessary.[5]

The Pregnancy Discrimination Act allowed women to continue their employment well into pregnancy and to return to work as soon as they were physically able. The intent of the law was to end the discrimination that pregnant women had historically faced in the workplace.

In the years following the passage of the PDA, several states developed policies mandating certain types of benefits for pregnancy and childbirth. California passed a law that guaranteed up to four months of unpaid leave for the purpose of child birth and recovery. The California Federal Savings and Loan Association challenged this law as a violation of the Pregnancy Discrimination Act, arguing that since the PDA stated that all employees should be treated the same, no special benefits could legally be allowed for pregnancy and childbirth.

The case went to the Supreme Court, and in 1987 the Court upheld the state of California's policy and clarified the intention of the Pregnancy Discrimination Act. The majority opinion stated that the law was intended to "construct a floor beneath which pregnancy benefits may not drop, not a ceiling above which they may not rise." [6] This decision provided the opportunity for individual employers to develop specialized types of maternity policies.

Some fire departments, as well as employers in other potentially hazardous work environments, developed policies for pregnancy in the workplace that were based on the concept of "fetal protection." Conditions and substances exist in many workplaces that offer either probable or proven hazards to a developing fetus and to the reproductive health of both men and women workers. These hazards include both chemical and environmental factors. Policies based on fetal protection often required a woman to leave a certain type of job, such as active firefighting, at some point in her pregnancy (in many cases, as soon as the pregnancy was known). This requirement was based not on the woman's ability or inability to perform her job, but rather on the employer's concern that some aspect of her work might prove harmful to the developing fetus.

The Supreme Court considered such fetal protection policies in its 1991 *UAW V. Johnson Controls decision*. This case concerned a battery manufacturer that had barred all fertile women from working in jobs using lead, based on the premise that lead could harm a fetus. The Supreme Court rejected this policy as a form of sex discrimination and a violation of the Pregnancy Discrimination Act. The majority opinion stated: "Decisions about the welfare of future children must be left to the parents who conceive, bear, support, and raise them rather than the employers who hire those parents." It further pointed out:

> It is no more appropriate for the courts than it is for individual employers to decide whether a woman's reproductive role is more important to herself and her family than her economic role. Congress has left this choice to the woman as hers to make. [7]

Although other court decisions relating to pregnancy and employment exist in the lower courts, no other Supreme Court decisions have modified or contradicted the thrust of *California Federal Savings & Loan* and *Johnson Controls*. These decisions provide guidance for policy development regarding pregnancy and maternity in the fire service.

Policy Development

Although employment policies that relate to pregnancy and childbirth are often grouped under the general heading of "maternity policies," there are really three types of policies that should be considered. These are: maternity leave policies, pregnancy policies, and parental leave policies.

Maternity leave policies. These policies address the specific period of time when a woman is disabled as a result of childbirth. All women who go through childbirth are temporarily disabled by it to some degree, whether it be for a few days or for several months. Allowing time for women to have their children and physically recover afterwards is a need in every workplace. When a woman's pregnancy does not restrict her ability to do her job, and when she suffers no complications from the pregnancy or delivery, a typical maternity leave might be six to eight weeks. Extensions may be available if the pregnancy and/or delivery are very difficult. In some cases, employers provide this time as paid or partly-paid leave specifically for childbirth. In other instances, women must use sick or vacation time to insure a paycheck during the time surrounding childbirth.

Pregnancy policies. Women firefighters need maternity leave just like all other women workers, but, like women in other hazardous professions, their needs are more complex than that. A typical maternity leave assumes that a pregnant woman can and will work at her usual job long into her pregnancy. The PDA guarantees women the right to do this as long as they are able to perform all of their job functions. However, in the case of active firefighting, there is real concern as to whether women should continue in their usual job assignments while pregnant.

A number of chemical and environmental factors can adversely affect pregnancy. *(See inset on medical considerations, page 45.)* Many of these substances or conditions have been shown to have adverse reproductive outcomes in both males and females in animal studies; for the most part, the human research as it relates to firefighting is too sparse to support solid conclusions. On the basis of the research that does exist, most physicians agree that women should stop fighting fires and doing other high-risk work at some point in their pregnancies. Exactly what point is most appropriate is a subject of some debate. Because many environmental hazards are most dangerous to a fetus during the first trimester (the first three months a woman is pregnant), many fire departments want women to leave hazardous duty as soon as their pregnancies are known. The majority of women prefer this arrangement as well. [8]

The general trend among career-level fire departments is to offer alternate duty to a firefighter during the term of her pregnancy. This type of assignment is sometimes known as "light duty." Volunteer departments may recommend that a pregnant firefighter limit her activities to non-hazardous duties within the department as well. Although a transfer to non-hazardous or "light" duty may be recommended during pregnancy, and may be supported by all employees, the Supreme Court decision in *Johnson Controls* strongly suggests that such transfers can not be mandatory. In response to the decision, some fire departments have changed policies of mandatory transfer to ones that strongly recommend such a change of assignment.

As research increases, more attention is being given to the reproductive risks of parental exposure

---

Medical considerations of firefighting and pregnancy

The job of firefighting presents many potential hazards to healthy reproduction. It poses physical hazards such as drastic temperature variations, extreme and unpredictable physical exertion demands, and psychological stress. Firefighters may also be exposed to biological or radiation hazards. Additionally, the fire environment may produce a wide range of chemical agents, including irritant and asphyxiant gases and other toxins.

Human reproductive health as it is affected by the work environment is a relatively new area of study. The clearest connectionbetween an environmental agent and adverse reproductive outcomes for both men and women is in the case of ionizing radiation, which is not a common hazard for most firefighters. Prolonged exposure to high ambient temperatures, however, may also have a detrimental effect on fertility and pregnancy. High heat exposure has been related to infertility in men and may be linked to neural defects in the babies of exposed mothers.

Chemical agents in the fire environment are numerous and unpredictable. The toxic effects of fire smoke have been tentatively linked to a number of physical problems, including respiratory disease, coronary artery disease and malignancies. Many chemical agents in the fire environment may also adversely affect reproduction. Carbon monoxide, carbon dioxide, hydrogen cyanide, acrolein and other aldehydes, sulfur dioxide, hydrogen chloride, nitrogen dioxide and benzene are all commonly produced in fire environments. Research shows that all of these compounds may have detrimental effects on reproduction. Pregnant women and their fetuses are especially affected by carbon monoxide exposures.

Although much more study is needed, existing research suggests that both men and women are vulnerable to reproductive toxicity in the firefighting environment. In addition, the potential hazards to developing fetuses pose special concerns for pregnant firefighters.

Source "Reproductive Hazards of Firefighting I and II," Melissa McDiarmid, M.D., et al., American Journal of Industrial Medicine, 19 433-472 (1991).

---

near the time of conception. Some people feel that women and men who are trying to conceive should avoid environments of potential risk to reproductive health. At least one fire department has implemented a policy of allowing, on request, a voluntary transfer to alternate, non-hazardous duty for any firefighter, male or female, who is trying to conceive.

The fire department must provide education if any reproductive safety policy is to be implemented successfully. All employees must understand the hazards to reproductive health that firefighting poses to men and women. A qualified physician, or other professional who is well versed in the existing research on the issue, should conduct classes on this subject for all firefighters and officers.

Women firefighters should also be educated early in their careers about the options that exist for them should they become pregnant. The fire department should develop reasonable policies and clearly define the options within them. The ideal policy is one where non-hazardous duty is guaranteed for the term of a pregnancy and while the mother is nursing. The transfer should be to

meaningful work that does not penalize the woman, and it should involve no loss of pay or benefits. Pregnant firefighters have worked successfully in such areas as training, public education, prevention and inspection, policy development and communications. Fire departments of all sizes have found ways to utilize the pregnant employee productively in a role that does not offer reproductive hazards; the department and the employee both benefit from the short-term assignment to a non-combat position. The International Association of Fire Fighters in 1992 adopted the position that the pregnant firefighter should be offered the opportunity for voluntary transfer from firefighting at any time during her pregnancy without loss of pay or benefits.

When considering options for alternate duty, the term "non-hazardous" offers a clearer description of what is being sought than does "light" duty. A light-duty assignment is not physically taxing, but it may still expose the worker to hazardous environments. For example, a light-duty task in one fire department required a pregnant firefighter to fuel all of the department's staff cars every day. Excessive exposure to gasoline and

Maternity and parental leave policy checklist

The specific terms of a fire department maternity policy will vary depending on the needs of the city or other employer. However, all comprehensive policies have some points in common. Keep the following guidelines in mind as you work to develop your department's policy.

The policy should guarantee that the pregnant employee will not lose her job. Firing an employee because of pregnancy is a violation of the Pregnancy Discrimination Act of 1978.

❏ Alternate, non-hazardous duty should be available but not mandatory. The employee who transfers to this type of assignment should not lose any pay or benefits by doing so.

❏ It is the employer's responsibility to educate all employees, both male and female, about reproductive hazards in the workplace, to enable them to make informed decisions about conceiving and bearing children.

❏ The employee's health care benefits should be maintained during any type of leave related to pregnancy.

❏ Pregnant employees should not be required to exhaust all sick leave, vacation time or other forms of personal leave, before being allowed to use maternity leave. Use of these other types of leave for the purpose of maternity should be at the employee's discretion.

❏ Pregnant employees should not lose seniority or eligibility for promotion due to any paid leaves or transfers to non-hazardous assignments.

❏ Unpaid leave should be an option, both during pregnancy and after the birth. Parental leave should be gender-neutral and equally available to both parents regardless of their marital status, as well as to non-biological parents.

❏ If maternity/parental leave is part of contract language, and the contract does not cover recruit and probationary firefighters, the department's policy should address what options are available to the recruit or probationary firefighter who becomes pregnant or becomes a parent.

diesel fumes during pregnancy may cause health problems. Similarly, doing arson investigation in freshly burned buildings is probably not the best choice for an alternate assignment. Non-hazardous duty will still permit the pregnant firefighter to take any training or recertification classes that other firefighters are undergoing, as long as the classes' activities do not pose any risks to her. This not only keeps the department from having to train her retroactively and keeps her from falling behind in her training, but it also allows her to remain in touch with her co-workers and with combat operations during the time she is reassigned.

A fire department is not required to offer alternate-duty assignments to pregnant firefighters unless it has a policy of reassigning all employees who have temporary disabilities. The essence of the Pregnancy Discrimination Act is that pregnant employees must be treated at least as well as other employees who are temporarily disabled. However, pregnant employees may be offered alternate duty even if that option is not available to all other employees. The legality of this type of variance in treatment was clearly upheld in *California Federal Savings & Loan.*

It is also important to remember that the law, as of 1992, upholds a woman's right to continue working in a hazardous environment even during pregnancy. The overwhelming majority of women will not choose to do this. However, if faced with a choice between a higher-risk pregnancy and economic devastation caused by a significant loss of pay and benefits during pregnancy (because no alternate duty is offered), some women will feel compelled to choose the former.

What is the best solution? Although pregnancy poses a challenge to fire departments, the issue is certainly manageable. Many cities and other jurisdictions have arrived at good solutions that consider the needs of both the employee and the employer. A policy that allows the employee to continue being a contributing member of the department during pregnancy, and that assures her of continued pay, benefits, and seniority at normal levels, is an attainable goal, even for small fire departments.

In a minority of cases, a pregnant employee will be unable to work even in an alternate duty assignment because of health complications during pregnancy. A leave of absence during pregnancy and childbirth should be an option on every fire department. Such leaves are usually unpaid, but may include continued accrual of seniority and benefits. At the very least, an unpaid leave should include continued medical insurance coverage, even if the employee must pay the full premium for the coverage. Unpaid leaves should be available for the duration of the pregnancy, and should include enough time to accommodate recovery from a delivery that involves complications.

Some fire departments have developed policies that require the opinion of a doctor, often the employee's personal physician, as the deciding factor for how long a pregnant woman can work safely in her fire service job. Since many physicians are not familiar with the actual demands and hazards of firefighting, they should be educated about the job before they are asked to give such an opinion. Fire departments may wish to develop a standard physician's release form for this purpose, which specifically lists the requirements of the job. Any such form used in cases of pregnancy must also be used in a comparable way for other non-duty-related disabilities, such as off-the-job injuries.

Parental Leave. Most women will be physically capable of returning to full firefighting duty within six to eight weeks following the birth of their babies. However, a new mother or father may wish to spend more time with an infant beyond the time needed for physical recovery. Increasingly, employers are offering what is known as parental leave for those employees who want or need more time to be full-time parents to a new baby or a newly adopted child.

Parental leave is not a new concept. Among the world's industrialized countries, the United States and South Africa are the only two that do not have national policies that guarantee leave for the purpose of parenting or family care. The parental leave policies of six European countries (Austria, Germany, France, Italy, Finland, and Sweden) guarantee from 12 to 52 weeks of leave, with 60-100% salary retained during all or part of the leave. Canada offers 17-41 weeks of parental leave, with 15 weeks at 60% salary guaranteed.

Many employers, including city or county governments and fire departments, have recognized parental leave as a policy that can benefit both workers and their employers. Parental leave policies may be negotiated into contracts, developed by managers as standard practices, or passed as local ordinances. In general, parental leave policies are gender-neutral: because parental leave is different from the type of leave needed for physical recovery from childbirth, either parent would be eligible for this type of leave. Four months of unpaid leave is a standard parental leave in this country, although some employers provide up to a year. In an increasing percentage of cases, such leave may also be used for elder care: employees who require time off work in order to care for an ill or bedridden older relative.

Legislation guaranteeing access to unpaid parental leave for most U.S. employees was passed by Congress in 1991 and 1992, but the legislation was vetoed both times by the President, who nonetheless stated that employers should offer this type of leave to their employees. The bill that was passed included care of a critically ill or injured family member as a valid reason for using parental (or "family") leave.

Fire departments have just begun to recognize the value of parental leave as a way of retaining good employees and of improving morale. Of fire departments from across the country surveyed in 1990, fewer than half provided some form of parental leave for all employees. One department allowed only one day of leave for a father when his child was born. The trend, however, is for fire departments to consider more realistic policies for parental leave and to extend these benefits to both partners involved in child-rearing, regardless of their gender.

Firefighters have unique requirements when it comes to child care. Most need something other than the normal day time care options that are commonly available. Firefighters' children often need non-parental child care for more than 24 hours at a time. Many firefighters are subject to emergency call-back during major incidents, and may need someone to take care of their children on a moment's notice at any hour of the day or night. Two-firefighter couples or single parents are especially affected by these circumstances. The problem particularly affects women firefighters: 31% of women firefighters are married to or involved with other firefighters; 11% are, or have been, single parents.

Many cities and private employers are beginning to take an interest in the child care needs of their employees. They recognize that child care problems cause absenteeism, poor productivity and morale, and may lead to the loss of good employees. Problems with child care are a significant reason why women may not enter the fire service, may leave early in their careers, or may not return after having children.

Although governments and private employers are giving more attention to this issue, most existing programs address only the needs of employees who work relatively conventional hours. The few employers' who support child care centers for their workers most often maintain these centers only during extended business hours. Centers that can accommodate 24-hour child care or emergency drop-ins at any hour are virtually non-existent. Other management practices that may help some employees with child care problems, such as job-sharing and flexible hours, are either not feasible or not often practiced for firefighters.

Creative solutions are needed and have begun to emerge. One fire chief has suggested that his city develop old fire stations into around-the-clock child care centers

> **Problems with child care are a significant reason women may not enter the fire service, may leave early in their careers, or may not return after having children.**

specifically for the benefit of firefighters' children. "We found that some qualified people do not apply for the job because they're concerned about what they'd do with their kids," he said. "I don't think it benefits the department if qualified people get away because of that." [10] (I In the U.K., the London Fire Brigade provides child care allowances to parents who pay babysitters, and subsidizes spots in child care centers for its employees' children. [11] The Suisun City, California, Fire Department has increased the off-duty response of its full-time personnel to major incidents by outfitting and staffing one of its rehab vehicles to handle child care. Firefighters responding to the call can drop off their children at the fire station or at staging, to be cared for by members of the rehab team for the duration of the incident.[12] Other fire departments provide child care during off-duty training sessions, to encourage greater participation. Positive examples are also being set by some hospitals and airlines, which have many employees who need child care at unusual hours.

As single-parent families and two-firefighter couples become more common, the issue of child care for firefighters will become increasingly important to fire service managers. Problems with child care can prevent individuals from entering the fire service or may influence their career tracks once they are there. The burden falls most heavily on women with children, especially single mothers.

Fire departments have historically had a strong family tradition. Sons have followed fathers into fire service careers for generations. Brothers have served side by side as career or volunteer firefighters. Now that women are an integral part of the fire service, a new family tradition has emerged. Although women may follow their fathers (or mothers) into the fire service, and sisters and brothers may both choose firefighting vocations, the most common type of familial relationship between male and female firefighters is that of marriage.

According to Women in the Fire Service's 1990 survey, 20% of all women firefighters surveyed were married to firefighters. Another eleven percent were involved in permanent or long-term relationships with other firefighters. Most of the marriages were between firefighters on the same department, and the majority of marriages and relationships developed after both employees were hired.

It should not be surprising that these figures are so high. Once a person has finished school, the workplace is the most likely place to meet friends, social partners, or mates. This is true in any profession, but seems especially so with firefighters. The nature of the job, particularly the unusual hours and work environment, can make a conventional social life difficult. Strong bonds of friendship and loyalty have always developed among firefighters who work closely together under difficult circumstances. It is natural that these same feelings would develop between men and women firefighters, and that good friendships might potentially lead to further involvement or commitment.

Fire departments have reacted to this trend of firefighter relationships in widely varying ways. Some have taken a completely laissez-faire attitude and have not interfered in any way with firefighter couples, even allowing them to continue working in the same station together. Other departments have taken the opposite approach, attempting to impose very restrictive policies on these employees. An extreme case of this type of policy was one that prohibited marriages between employees and made marriage to another firefighter a cause for dismissal. The policy was made effective retroactively so that the one woman on the department (who was married to a co-worker) would be affected. Another example would be of departments that attempt to discover which couples are dating or otherwise emotionally involved with each other, and reassign one or both parties. Such policies go beyond legitimate employer concerns into the realm of harassment, resulting in unnecessary stress for the employees as they attempt to conceal relationships and even marriages from their co-workers.

Policies against nepotism (favoritism to a relative) are a legacy of 19th-century political "spoils" systems and were originally put in place to prevent local political officials from appointing their relatives to jobs. They are stronger in some parts of the country than others; of fire departments recently surveyed, less than a third had a formal written anti-nepotism policy. Some policies included strong statements such as the following:

> The employment of relatives in the same organization tends to have a number of undesirable results. In the interest of preventing potential abuses in hiring, supervisory authority, or the appearance thereof, it is the policy of the City to limit hiring and supervision of relatives by City employees.

Others took a more moderate approach:

> The department recognizes that there are many situations where two individuals who have a personal relationship may appropriately be allowed to work in the same program, activity or location without adverse impact. However, under circumstances where work, safety, morale, or impartial supervision is demonstrably and adversely impacted by a personal relationship, the affected employees may be accommodated by the reassignment of one or the other.

Some anti-nepotism policies apply to all employees of a city or county government, not just the fire department. These policies may not be enforced at all in some cases, or may be enforced unevenly from department to department or individual to individual. Such inconsistency in the enforcement of a policy can create exactly the type of problems that the policy is designed to prevent.

The anti-nepotism policies of some cities restrict the hiring of relatives but do not address the potential for employees to become related through marriage once both individuals are employed. Restrictive policies written specifically to address couples on the job may require the transfer or career limitation of one or both employeesaffected. Since women firefighters often have less seniority and lower rank than their partners, such policies may have a disparate, negative impact on women employees.

Relationships in the workplace are a fact of life. They need not be viewed as a problem to be solved or as a situation to be prevented. Where real problems do exist, either with the couple themselves or due to co-workers' or managers' resentment of the couple, they should be dealt with individually. Perhaps the best policies are ones that take into account job performance as they seek to manage relationships in the work environment. It is possible to develop policies that both respect the rights and privacy of individuals and maintain a professional work environment for all employees.

The issue of hair standards continues to be unresolved for firefighters of both genders, There are several reasons for the continuing controversy as it relates to women. Many departments are still hiring their first women and so are just going through the process of addressing their hair standards. It also appears that as women become more a part of the fire service, both new and experienced women firefighters are becoming less willing to sacrifice a favored hair style simply to comply with standards that were designed for a male workforce. As the fire service moves away from a paramilitary organizational approach to its employees, fire service managers in general are adopting a more flexible approach to employee grooming standards.

A fire chief who takes an extreme approach, such as requiring all employees to have military-style short hair based on male norms, is demonstrating the department's hostility to the presence of women and may be risking a sex discrimination charge since, based on traditional hair styles, most women would be excluded from the job while most men would not. The search for a clear answer is made more difficult by discrepancies in the way the courts and the EEOC have decided hair-rule challenges; these discrepancies have contributed to the confusion that surrounds the issue.

This section of the manual will explore the legal background for firefighter hair-length requirements, from the perspective of both the EEOC and the courts. It will discuss the two primary reasons for fire departments to have hair-length policies, and offer guidance in policy development. Although this section does not explicitly address related grooming issues such as the wearing of jewelry, some references to jewelry are provided in the sample policy language on page 51.

The legal background

The courts have generally allowed employers more latitude in establishing work rules for hair length than they have in dealing with "physically immutable" or unchangeable characteristics such as height. Many early cases were brought by men, including police officers and firefighters, challenging standards which prohibited long hair on men. Since the mid-1970's, the courts have consistently held that prohibiting long hair for male employees is not sex discrimination under Title VII if the employer's grooming standards for both sexes are related to community standards and are applied in an even-handed manner. [13] The courts have made a distinction between hair standards and issues where discrimination affects "fundamental rights" or is based on immutable characteristics."'

The judicial and administrative branches of the government disagree on a basic point of sex discrimination law as it affects hair and appearance standards on the job. The courts have defined Title VII as a weapon against sex-linked practices that seriously impair fair employment, but have decided that the law was not intended to interfere with employers who wish to set reasonable appearance standards required by their business. The Federal Equal Employment Opportunity Commission (EEOC), on the other hand, has consistently treated different appearance rules for men and women as constituting sex discrimination under Title VII.[15] At the same time, the EEOC has upheld reasonable safety hazard concerns in industry.

While the applicability of Title VII to hair standards is thus undecided, the constitutional protections afforded to employees in the workplace are clearly limited. Numerous cases have been brought on constitutional grounds. In these cases, employees have argued that appearance is an aspect of personal liberty, so that any interference with that "right" must be justified by a legitimate state interest. The courts, in response, have generally upheld a public employer's right to impose grooming regulations that can be justified by the need for discipline, uniformity and *esprit de corps,* even where no safety considerations exist.[16]

However, a 1992 case filed under Louisiana constitutional law resulted in a ruling against a public employer, the City of Shreveport. At issue was the city's attempt to restrict firefighters' hair to shoulder length or shorter. (Women had been on the fire department for eleven years and had previously been permitted to pin their hair up to conform with the hair standard.) In issuing an injunction against the employer, the judge found that "the regulation (was) not gender neutral" because it affected women differently from men and "ha(d) the effect of classification of individuals on the basis of sex." The ruling continued:

> Similarity of appearance can be and was in fact achieved by requiring fire personnel on duty or in uniform to have their hair, if longer than regulation length, pinned to meet a length requirement established by their employer. The esprit de corps and equal treatment as togrooming standards are easily achieved by uniform enforcement of hair length *while on duty or in uniform* with a recognition of an individual's right to have whatever length of hair he or she desires *as long as while on duty or in uniform it is kept to a level set forth in the employer's regulation...* While similarity of appearance has been recognized as an appropriate and rational goal in a "para-military" civilian service, commonality may not and should not be required at the expense of reason and purpose."

Safety vs. grooming

Apart from the legal aspects, confusion also exists on hair-length issues because fire departments have, in the past, required firefighters to have short hair for two different reasons: because it was, with the prevailing protective gear, safer during firefighting operations, and because it looked neater, more professional, or more uniform. Where these two reasons have become intertwined into one policy, it is necessary to sort out the real and justifiable reasons for the policy.

A safety standard is gender-neutral: fire will burn exposed hair regardless of the sex of the person whose head it's on. Advances in protective equipment in recent years have made it possible for firefighters to have much longer hair than in the past and still be much safer than ever. SCBA facepieces and flame-resistant hoods and helmet liners insure that all surfaces on the head are protected. For this reason, many fire departments have adopted a safety-based hair standard that simply states that "No hair shall be exposed during firefighting activities."This approach avoids the need to regulate the actual length of the hair, and bases its restriction simply on the justifiable need for firefighter safety.*

Establishing a policy that is based solely on grooming and appearance concerns has a more favorable impact on women, for while a safety standard must be gender-neutral, a grooming standard need not be. Even though the EEOC disagrees, the courts have consistently held that employers can legally require male employees to wear their hair shorter than female employees and that such policies do not constitute illegal sex discrimination.

*In the mid-1980's. the Madison, Wisconsin. Fire Department administration, which had previously had a policy requiring extremely short hair for both men and women firefighters. reassessed the need for such a policy. In doing so, it determined that long hair, if not exposed during firefighting operations, was no more dangerous than short hair, and in fact the bulk of hair pinned up inside the helmet actually offered insulating protection to the firefighter's head. The policy was changed to a "safety only" standard.

Examples of language from fire department hair-length and grooming policies

"Hair shall be neatly groomed and the length or bulk of the hair shall not be excessive or present a ragged, unkempt or extreme appearance. *(Men:)* Hair may cover one half of the ear but shall not cover the entire ear. (Women:) Hair may not extend beyond the lower part of the shoulder blades.

"Members are discouraged from wearing rings or other jewelry on the fire or training ground. Female members may wear earrings providing they do not extend below the bottom of the ear."

(Women:) "Hair must be clean and neatly arranged. When in uniform, back hair must not fall more than one-quarter inch below the lower edge of the collar. No hair must show under the front brim of fire service headgear. Afro, natural, bouffant, and other similar hair styles are permitted, but... bulk of hair must not exceed two inches. In no case is the bulk of the hair permitted interference with the proper wearing of fire service headgear.

"Only pins, combs or barrettes that are similar in color to the individual's hair color may be worn to meet the requirements of the regulation. Jewelry which extends beyond the ear lobe or... is loose or protrudes and may catch in machinery or equipment may not be worn while on duty."

"It is recognized that traditionally acceptable standards for female hairstyles differ considerably from those of males. Female hairstyles that would normally not conform to the standards outlined in this S.O.P. may be pinned up or secured in order to comply while on duty. In these instances, the hair must be pinned up or secured at all times while on duty, and shall not interfere with the proper wearing of uniform hats or protective equipment, or in any way create a safety hazard."

(Men and women:) "There are many hair styles that are acceptable. So long as the person's hair is kept in a neat, clean manner, the acceptability of the style will be judged by these criteria: Hair styles that preclude the proper wearing (of SCBA) are not permitted... Hair will be worn so that it does not extend below the bottom of the uniform shirt collar while standing in an erect position. Hair may be pinned or worn in a way to keep hair above the bottom of the collar..."

"To facilitate a professional appearance, hair and grooming standards must be followed. These standards havebeen modified to meet contemporary styles without jeopardizing the safety of firefighters involved in the hazardous activities associated with firefighting.

"When in a normal standing position, the hair can extend to the top of the collar area. Hair will not extend beyond the bottom of the earlobe. Longer hair is acceptable if it is pinned up in a neat manner and does not interfere with the wearing of departmental headgear. No ribbons or ornaments shall be worn in the hair except for neat inconspicuous bobby pins or conservative barrettes which blend with the hair color. Hair... will not exceed two inches in height."

Policy development

Firefighter hair length can easily become a controversial issue within a department. From a common-sense perspective, even if not a legal one, finding a solution requires a balancing of employee and employer interests. For the employee, hairstyle is a matter of personal identity, and a preferred hair length can not be put back on at the end of a work shift in the way that a uniform is removed. The employer's responsibility for personnel safety and their interest in a professional appearance can encompass a wide diversity of hair lengths and styles. The best grooming standard may well be one that screens both sexes on a community-based standard of dress and appearance, a standard that applies an equal burden to both sexes.

Most fire stations in use today were planned and built with a single-gender workforce in mind. Many of these buildings are now being used by a workforce that includes both women and men. Not surprisingly, this can result in inadequacies that are a source of inconvenience, discomfort, embarrassment, and friction for all concerned.

Different fire departments have developed a variety of solutions to the problems created by inadequate facilities. The cheapest and easiest answers are usually the first to be implemented: a "men/women" or "occupied/unoccupied" flip sign on the door of the station's only restroom or shower, or a lock on the door, can be readily installed. Makeshift partitions, such as a row of lockers or a rollaway curtain, can often be put up between beds if bunkroom separation is desired.

These are short-term solutions to real or perceived needs concerning personal privacy in the fire station. The underlying question that guides the development of long-range answers is whether men and women on the job should be provided with separate facilities or not.

The legal background

The expense to the employer of providing separate restrooms, showers, locker rooms and bunkrooms for women and men usually will not support excluding women from an occupation.[s] According to the EEOC Guidelines,

> Some states require that separate restrooms be provided for employees of each sex. An employer will be deemed to have engaged in unlawful employment practice if it refuses to hire or otherwise adversely affects the employment opportunities of applicants or employees in order to avoid the provision of such restrooms for persons of that sex. [19]

An even more emphatic provision occurs in the Guidelines of the Office of Federal Contract Compliance:

> The employer's policies and practices must assure appropriate physical facilities to both sexes. The employer may not refuse to hire men or women, or deny men or women a particular job because there are no restroom or associated facilities, unless the employer is able to show that the construction of the facilities would be unreasonable for such reasons as excessive expense or lack of space. [20]

In one case where a firm had refused to hire a woman welder on the grounds that its repair yard lacked locker and restroom facilities for women, the EEOC discovered that the employer had actually had separate facilities during World War II when it had many women workers. [21] Since the existence of male-only facilities is often the result of the past discrimination that Title VII was designed to eliminate, allowing cost as a defense would only honor and perpetuate that discrimination.

One state law that applies to some fire stations is Section 2350 of the California Labor Code. It requires that business establishments that have five or more employees must provide separate bathrooms for each sex, and that no person may use bathrooms designated for the opposite sex. An interesting question that has come up regarding interpretation of this law is whether a fire station that normally houses a crew of four could be considered at shift change to "have" eight employees, particularly if it is common practice for the oncoming crew members to arrive early in order to take showers and change their clothes. Other states may have comparable provisions in their labor codes or other laws.

Local building or health codes usually require employers to provide bathrooms, and sometimes other facilities, for each gender in the workplace. As most fire stations are the property of municipal or county government, they have generally been made exempt from the provisions of these codes. Where such exemptions do not exist, of course, the fire department would be responsible for compliance.

A few fire departments assign women only to stations that "have facilities for women." This is not generally considered an acceptable long-term solution, particularly where there is an appreciable difference in the kind of work carried out at the different stations - a lighter or heavier call load, different types of apparatus, specialty teams-or where station assignments are on a seniority bid basis and the woman would otherwise be entitled to bid for the off-limits stations. The result can be an unworkable inflexibility for management, particularly if it prevents women from "roving" or "floating" to certain stations for one or a few shifts. It can also generate resentment from male co-workers that the woman doesn't have to take her turn at roving, particularly if her low seniority would normally make this part of her job. And it is unfairly restrictive of the woman, if she is not permitted to make shift exchanges or time trades with firefighters at certain stations.

The impact of inadequate facilities

One fire service observer, commenting on the problem of inadequate station facilities, wrote:

> Under the best circumstances, bad facilities are an inconvenience which women suffer from in far greater proportion. Under the worst conditions, poor facilities can lead to problems with morale and job performance, and an increase in the occurrence of harassment. At least one discrimination lawsuit has been filed which was due in part to inadequate facilities. A lawsuit costs a lot more than a locker room, and in the end, no one wins.
>
> When the need for women's facilities in the fire station is neither recognized nor addressed, the...

department may be saying that women are not important enough here to deserve decent facilities, that women may not be around long enough to warrant planning for the future, that women are not wanted at this station, and this is a reasonable way to keep them out; or that we are too busy here to consider the real needs of our personnel. All of these are harmful messages, both for women and for the organization of which they are a part. [22]

There are two common effects of forcing firefighters in a newly integrated workforce to occupy inadequate facilities. One is that the women are blamed for "causing" the problem. Even though it is the design of the station that is lacking, the feeling among the men is often that since "there wasn't a problem until she got here," it's the woman firefighter's fault. Solutions such as bumping an officer out of his private room to give it to the woman can generate a similar resentment. Where this type of hostility exists, providing facilities that offer privacy for both genders becomes only half of the solution. It is important for management to make it clear that alterations to the facilities are being done not "for the women" but for better privacy for women and men alike.

> **Women and men in the workforce - and particularly women, if they are in the minority or are the newest firefighters - will usually adapt to situations that are less than ideal... But just because a person... can develop behaviors to cope with a situation or environment doesn't mean it's right to leave things that way indefinitely.**

Another common reaction to inadequate station facilities is the tendency to adapt and accommodate to them. Women and men in the workforce - and particularly women, if they are in the minority or are the newest firefighters - will usually adapt to situations that are less than ideal. Many women firefighters do not routinely shower at work, or they get up an extra hour early in the morning in order to shower before the men need the shower facilities. Women have learned to use broom closets as changing rooms; firefighters of both genders develop the habit of looking for feet under the restroom stall walls. But just because a person or group can develop behaviors to cope with a situation or environment doesn't mean it's right to leave things that way indefinitely. All firefighters deserve some basic privacy, either individually or by gender, for personal functions. Management should make it a priority to provide this.

Possible solutions

Develop a five- or ten-year plan for remodeling your firehouses. All new stations that are built, and any significant remodeling of existing stations, should include adequate facilities for a two-gender workforce. Women may not be on the department now, but most fire stations will stand for a half century or more. To continue to build them on old designs means you will only be creating more of the same type of problem that firehouse designers fifty years ago created for you today.

Although crews in many fire stations manage to cope quite well with shared facilities, it is preferable for station design to provide privacy for both sexes in restroom, shower, and changing areas. The issue of separate bunkrooms for women and men is more controversial. As mentioned earlier, taking an officer's room away to give to a woman firefighter usually creates hard feelings. Tucking an ad hoc "women's bunkroom" off in one corner of the station (such as a rollaway bed in the weight room) is inconvenient for everyone and a clear message that the woman doesn't belong. The most common solution is for women and men to share the one existing bunkroom. Many women firefighters prefer this arrangement, because it keeps them a part of the crew and a part of the information-sharing process that begins as soon as a call comes in. On the other hand, some men and women are not at all comfortable sharing a bunkroom in this fashion.

The real long-term solution to the bunkroom question is to provide privacy for everyone. Many new firehouses are now being built to a design that features cubicles containing a bed, desk, lamp, and three or four lockers (for one person on each shift), with a curtain across the doorway. This provides privacy and a reduction of sound or light from the others in the room. (Snoring may be a common source of humor among firefighters, but routinely being deprived of sleep by one or more snoring co-workers is also a significant source of job-related stress.) If the partitions do not extend all the way to the ceiling, the open space at the top allows for air circulation and allows emergency tones and information to be heard.

This is a solution that pleases everyone and doesn't pit the women against the men, the paramedics against the suppression personnel, or the officers against the firefighters. It also avoids controversies over whether the women's bunkroom in the new station should be the same size as the men's bunkroom, whether men are allowed to use the women's bunkroom if no women are assigned to the station on that shift, whether a station that houses two officers should have men and women officer's bunkrooms, etc. It is a solution that respects the privacy and individuality of all firefighters without regard for gender, and for that reason is usually supported by all concerned.

Notes:

[1] *New York Times,* January 17, 1986.

[2] Willing, Linda, "Maternity Survey," *Women in the Fire Service Quarterly.* Summer 1987, pp. 1-6.

[3] Women in the Fire Service, 1990.

[4] 42 U.S.C. §2000e (K)

[5] *See EEOC Questions and Answers on Pregnancy Discrimination,* C.C.H. ¶3951 (April 20, 1979).

[6] *California Federal Swings & Loan v. Guerra,* 107 S.Ct. 683,42 F.E.P. 1073 (1987).

[7] Supreme Court of the United States, No. 89- 1215. *UAW v.. Johnson Controls Inc..* majority Court opinion delivered by Justice Blackmun on March 20, 1991.

[8] "Those surveyed were asked to describe... the ideal maternity policy. The vast majority outlined a policy that would provide light duty for the term of the pregnancy... then leave permitted during actual delivery and for three to six months after birth." Willing, Linda, "Maternity Survey." *Women in the Fire Service Quarter/y,* Summer 1987, p. 4.

[9] Only two percent of all government and private employers sponsor day care centers for their employees' children, according to the U.S. Labor Department Bureau of Statistics (1988).

[10] Bowers, Karen, "Old firehouses may get rebirth," *Rocky Mountain* News, 8/l8/89: quoting Denver Fire Chief Richard Gonzalez.

[11] Allcock, Ann, "Workplace Child Care and the London Fire Brigade," *Women in the Fire Service Quurterly,* Summer 1991, pp. 8-9.

[12] Stevens, Larry H., "But who will watch the kids?" *Fire Chief* April 1992, pp. 113- 114.

[13] *Dodge v. Giant Food, Inc.,* 488 F.2d 1333 (D.C. Cir. 1973); *Baker v. Calif. Lund Title Co.,* 507 F.2d 895 (9th Cir. *1974) cert denied,* 422 U.S. 1046 (1975); *Longo v. Carlisle DeCoppet & Co.,* 537 F.2d 685 (2d Cir. 1976); *Earwood v. Continental Southeastern Lines, Inc.,* 539 F.2d 1349 (4th Cir. 1976); *Barker v. Taft Broadcasting Co.,* 549 F.2d 400 (6th Cir. 1977); *Knott v. Missouri Pac. R.R. Co.,* 527 F.2d 1249 (8th Cir. 1975).

[14] E.g., an employer's refusal to hire women with small children *[Phillips v. Martin-Marietta Corp.,* 400 U.S. 542 (1971)] or the firing of women who marry *[Sprogis v. United Airlines.* 444 F.2d 1194 (7th Cir. 1971)].

[15] E.g.. EEOC Dec. No. 71-1529, 3 F.E.P. 952 (5/9/71).

[16] *Quinn v. Muscare,* 425 U.S. 560, *reh'g denied,* 426 U.S. 954 (1976) [male firefighter challenge to hair standard]; *Kelley v. Johnson,* 425 U.S. 238 (1976) [male police officer].

[17] *Sellers v. Citv of Shreveport, et* al., 1992. The employer defended the policy as a grooming standard, stipulating that the safety of firefighters was not at issue.

[18] As a rule, expense will not support gender as a bona fide occupational qualification unless the expense would be clearly unreasonable. See EEOC Case No. YNY 9-047 (5/21/69), C.C.H. Empl. Prac. Guide ¶6010.

[19] 29 C.F.R. §1604.2(b)(5).

[20] 4 I C.F.R. §60-20.3(e)(1970).

[21] EEOC Dec. No. 70-558 (2/l9/70), C.C.H. Empl. Prac. Guide ¶6137.

[22] Willing, Linda, "Bedrooms and Bathrooms: The Hidden Message," *WFS Quarterly,* Winter 1988-89, pp. 1-2.

Sexual harassment is a form of sex discrimination. It is illegal, it can devastate those who experience it, and it often destroys the morale and productivity of the work environment. It is also widespread in the fire service: at least 60-74% of women firefighters have experienced some form of sexual harassment at work or in their volunteer departments, from relatively minor incidents up to sexual assault and rape.[1]

Sex discrimination on the job occurs when employment decisions are based on an employee's sex instead of his or her qualifications. Sexual harassment occurs when an employee receives unwelcome behavior from the employer (or someone under the employer's control) that only happens because of the employee's sex. There are two types of sexual harassment. *"Quid pro quo"* harassment involves demands for sexual favors in return for employment benefits. "Hostile environment" harassment involves behavior motivated by the target's gender that makes the workplace offensive, hostile or intimidating. *(See inset below for examples.)*

The key concept in defining and identifying sexual harassment is that it is unwelcome. Sexual harassment is not "natural," "normal," or a "compliment." It is a power play by the harasser that is degrading and humiliating to the target. The common motivation for sexual harassment is aggression, not sexual desire. The physical appearance and behavior of the target are not the "cause" of sexual harassment. Sexual harassment of all kinds is increasingly viewed as a form of violence against women. (The vast majority of harassers are men, and most victims of harassment are women.)

Even though sexual harassment is a common occurrence, it is widely misunderstood. Persistent myths that tend to trivialize it include the idea that women secretly enjoy the sexual attention, and that a woman who brings sexual harassment charges is probably lying to get back at someone. Although there are legal sanctions and, usually, employer policies against it, sexual harassment is grossly under-reported. Studies consistently indicate that only in about 5% of cases does the victim (or "target") report an incident of harassment. Rather than come forward with a complaint, women are much more likely to leave a job, request a transfer, or suffer in silence and hope the problem goes away.

> **Only in about 5% of cases does the victim... report an incident of harassment. Rather than come forward with a complaint, women are much more likely to leave a job, request a transfer, or suffer in silence and hope the problem goes away.**

Examples of *"quid pro quo"* harassment:

"If you'll go out with me tomorrow night, I'll make sure you get all the training you need to make your probation."

'You're not driving the engine unless you give me a kiss."

"I'll be on the interview board for your promotion. If you expect me to give you a good rating, you'd better sleep with me."

Examples of "hostile environment" harassment:

Unwanted touches, kisses, and other personal/sexual attention
Referring to women in derogatory, vulgar, and/or sexual terms
Posting of graphic or violent pornography in the workplace

The tendency for women not to report harassment is just as strong in the fire service as it is in other workplaces. Following are representative comments from women firefighters who have been harassed but not reported it:

> I didn't want to be labeled a troublemaker, and I didn't feel a positive outcome was possible.

> Much of the harassment occurred when I was on probation and felt I could not speak out... I did attempt to speak to my officers; all of them shrugged off my appeals for help.

> I didn't want to be "singled out" even more.

> I want to keep my job. It's clear that those who seek legal recourse can't come back to work.

> I overlook most of it and just keep on going. [2]

Women's reluctance to report incidents of sexual harassment stems from several sources. First, the interactions involved are very personal and of a nature that many individuals are not comfortable discussing. Second, a woman's socialization encourages her to feel sorry for the harasser or to believe that something in her own behavior caused the harassment. Third, targets of harassment may sense-often with justification -that complaining would do no good and might even make matters worse. Sexual harassment is an intimate violation that often occurs without witnesses. Women who are its targets generally feel powerless to stop it. Women firefighters, who are usually subjected to strong pressures to "go along to get along," and to fit in and be "one of the guys," are especially deprived of most forms of support in trying to put a stop to workplace harassment.

People who are targets of sexual harassment exhibit a wide range of responses to the behavior. Characteristically, most do not feel that many options are open to them, and they therefore may respond in ways that are not effective in getting the behavior to stop and do not help management learn that a problem exists. Typical responses fall into several recognizable categories.

> Detachment: joking, removing oneself from the seriousness of the problem
>
> Denial: pretending the problem doesn't exist
>
> Relabeling: excusing the behavior, telling oneself the harasser "was just trying to be friendly"
>
> Appeasement: trying to placate the harasser
>
> Self-blame
>
> Endurance: putting up with the situation
>
> Avoidance: quitting one's job or requesting a transfer
>
> Social support: counseling, seeking out others who have been harassed
>
> Confrontation: informing the harasser that their behavior is unwelcome
>
> Reporting [3]

The law on sexual harassment

The EEOC Guidelines define illegal sexual harassment as follows.

Unwelcome sexual advances, requests for sexual favors, and other verbal or physical conduct of a sexual nature constitute sexual harassment when:

(1) submission to such conduct is made either explicitly or implicitly a term or condition of an individual's employment,

(2) submission to or rejection of such conduct by an individual is used as the basis for employment decisions affecting such individual, or

(3) such conduct has the purpose or effect of unreasonably interfering with an individual's work performance or creating an intimidating, hostile, or offensive working environment.

[Source: 29 C.F.R. §1604.11.]

Denying that sexual harassment is a problem simply because women fail to report it, ignores the ways in which women are forced to submit to harassment or deny that it is happening to them. Given these pressures, a target of harassment cannot be said to be "tolerating" or "voluntarily submitting" to the behavior if she fails to complain about each and every incident. Women may be reluctant to report sexual harassment for a number of reasons.

> Lack of confidence in the system: believing the employer will do nothing to stop the problem
>
> Fear of retaliation: fearing the harassment will get worse, or the employer will retaliate in some way
>
> Fear of loss of privacy: being concerned that the target's personal life and past will be dragged in as somehow relevant to the harasser's behavior
>
> Isolation: lack of support systems for reporting in the first place; loss of any friends or allies in the workplace if the target "rocks the boat"
>
> Embarrassment
>
> Financial need
>
> Sense of guilt: unwillingness to get an otherwise "nice guy" in trouble
>
> Socialization

Women's unwillingness to report incidences of sexual harassment is often justified in view of the "victim-blaming" that can occur when complaints are made, and of the types of retaliation that often follow.

Sexual harassment is illegal under federal law,[4] state, local or municipal fair employment laws, state tort laws,* and some state unemployment and workers' compensation laws. It is also a violation of some collective bargaining agreements. The first court case to define sexual harassment as a form of discrimination was decided in 1976.[5] The federal Equal Employment Opportunity Commission (EEOC) first issued guidelines on sexual harassment in 1980.[6] Another landmark was established in 1986, when a unanimous Supreme Court affirmed that employees have a right to a work environment that is free of sexual harassment.[7] Noting that Title VII forbids discrimination in "terms, conditions and privileges of employment," the Court agreed with the EEOC that Congress intended, through Title VII, "to strike at the entire spectrum of disparate treatment of men and women in employment,"* including non-economic injury ("hostile environment"). [Title VII of the Civil Rights Act of 1964, as amended, is the federal law banning discrimination on the basis of race, religion, sex and national origin.]

Some courts have taken the position that if the harasser makes sexual overtures to or engages in harassing conduct with employees of both sexes, or if the conduct is equally offensive to men and women workers, the plaintiff does not have a remedy under Title VII. Even in such cases, however, the plaintiff might still have a claim under state or local laws, depending on the jurisdiction.

Hostile work environment harassment is particularly a problem for women who enter a predominantly male field such as firefighting because rough language, sexual jokes, "centerfold" postings, sex-based videotapes or television programming such as X-rated cable channels, touching, gesturing, and frequent comments about clothing, body, lifestyle and behavior are excused as "normal" hazing or treatment as "one of the boys." Some courts have found in favor of employers whose workplace is "permeated by an extensive amount of lewd and vulgar conversation and conduct,"[9] if the complainant initiated and participated in the conduct. This perspective, however, fails to take into account the social pressures on women in non-traditional jobs to become accepted by adapting to the highly sexualized male cultural norms of the workplace.

Subjectivity of offensiveness: the "reasonable person" and "reasonable woman" standards

Helpful though it would be, it is impossible to make a comprehensive list of all behavior that could constitute sexual harassment. Many situations must be examined individually; what is harassment in one case may not be in others. This can seem confusing until one focuses on the key concepts: "unwelcome," "intimidating or hostile," and "interfering with work performance." What is important is the effect of the questionable behavior on its target or, in some cases, on others in the work environment.

Only "unwelcome" conduct can be considered sexual harassment, but behavior that is amusing to one person may be unwelcome to another. In cases that go to court, the court must determine whether the complainant was actually offended or oppressed by the conduct in question. The court will also look at whether the complainant's reaction to the offensive behavior was "reasonable." Courts are still struggling with what standard determines whether the complainant's reaction to arguably abusive conduct was "reasonable." Some courts have adopted a "reasonable person" standard: the harasser's conduct must be such that it would affect the work performance and psychological well-being of a reasonable individual, as well as being actually offensive to the particular complainant. Not surprisingly, the social attitudes of particular judges have played a large part in determining what the reasonable person is expected to endure. However, many people expect this to change dramatically as these cases begin to be heard by juries pursuant to the Civil Rights Act of 1991. *[See Appendix 1 for a summary of the 1991 Civil Rights Act.]*

Recent legal factors affecting sexual harassment litigation

Many observers expect the passage of the 1991 Civil Rights Act to give a significant boost to sexual harassment lawsuits. For the first time, the law permits a jury to hear these allegations. Juries, compared to the judges who have heard these cases to date, may be more likely to display shock and outrage over harassment. This may translate into damages: also for the first time, the law permits compensatory and punitive damages to be awarded in sexual harassment cases.[10]

Another boost to sexual harassment lawsuits also came in 1991 when a federal judge approved the first class-action lawsuit for sexual harassment." Previously, sexual harassment suits have only been brought by individuals. Class action lawsuits have a greater impact and permit harassment victims to pool their resources. The *Eveleth* case was brought by three women iron miners and claims that all 100 women who worked or applied for work at the mine were subjected not only to discrimination in hiring and in terms and conditions of employment, but also to a pattern of sexual harassment and a hostile work environment.

---

*Applicable laws might include those prohibiting intentional infliction of emotional distress, wrongful discharge, contract interference, invasion of privacy, and assault and battery.

Stopping sexual harassment: the fire chief's role

Fire chiefs of career, volunteer, and paid-on-call departments who wish to minimize problems of sexual harassment should consider the following steps.

❑ Adopt written policies prohibiting sexual discrimination and sexual harassment, and a workable, confidential, step-by-step procedure for the filing of complaints.

❑ Provide training for all personnel - firefighters, officers, and support staff - on multicultural issues, including sexual harassment. Anti-woman behavior is part of the dominant culture of many firehouses and should be addressed as such. These sessions should facilitate discussion, and not simply consist of reading the department's policy and considering everyone to have been "trained."

❑ Remember that "what you permit, you promote." Do not ignore harassment. To do so sends the message that you are in agreement with the harassing behavior or discriminatory attitudes. Do not place all of the burden for reporting and correcting the problem on the harassed individual or group. Each stage of prejudiced behavior encourages the next; extreme behavior develops when more subtle behavior is permitted to continue.

❑ Support - do not discourage - people who bring complaints of harassment to your attention. Handle complaints confidentially and in a timely manner. An open-door policy goes a long way towards resolving problems at the lowest level, long before the chief is faced with a polarized situation or an expensive lawsuit.

❑ Prevent retaliation against those who have filed complaints, and against any witnesses who support their allegations. If the charge is found to have merit, discipline the harasser appropriately. Don't "solve the problem" by transferring the target to a new station against his/her will.

❑ Ensure personal privacy for everyone in station accommodations. Lack of privacy increases the tensions and resentment that lead to harassment.

❑ Keep in mind that just because you do not hear about sexual harassment occurring does not mean that it is not there. Use exit interviews (interviews with people who leave the job) as a reality check. A good manager will want to understand the reasons behind an employee's request for a change in status, whether it is an application for transfer or a notice of resignation.

❑ Be a role model. Avoid patronizing behavior or tokenism toward any group; be aware of language, policies, images, or situations that have questionable gender or racial connotations (e.g., the term "non-white") and correct any stereotyping that occurs. Be aware of, and correct, behavior that reflects your own prejudices.

The Sixth Circuit has taken the position that a woman who "voluntarily" enters a workplace must take it as she finds it.[12] The woman's workplace in that case included pictures of nude women, and a co-worker who routinely referred to women by derogatory sexual epithets. The court held that the obscene language might be "annoying" but not "startling" to the reasonable person because society condones, openly displays and commercially exploits erotica in public places such as newsstands, television and the movies. One judge, in a partial dissent, criticized the majority's "what you see is what you get" attitude. The dissent pointed out that the "reasonable person" standard fails to take into account the divergence between women's and men's views of appropriate conduct.[13] (For example, in one study, 67% of men surveyed said they would be complimented if they were propositioned by a woman at work. When women were asked if they would take such a proposition from a man in the workplace as a compliment, only 17% said yes.[14]

The Sixth Circuit decision failed to recognize the difference between pornography encountered in private life and in the workplace. Such material on a home television, in the movies or on the newsstand can be turned off, not attended, or avoided. Erotica and pornography in the workplace are usually displayed so that they cannot be avoided. Further, the not-so-subtle message of the material, and of its presence being condoned in the workplace, is that the sexual objectification and trivialization of women workers is acceptable.

Other courts, recognizing that a woman's perspective may differ substantially from that of a man, have adopted the "reasonable woman" standard. The Ninth Circuit, and district courts throughout the country, have concluded that pornographic posters and language could seriously affect the psychological well-being of the reasonable woman and interfere with her ability to perform her job.[15] In a case involving nude calendars posted at the workplace, the court stated:

> The proliferation of [pornographic] material may be found to create an atmosphere in which women are viewed as men's sexual playthings rather than as their equal co-workers.[16]

Less clear-cut from a legal standpoint are calendars, posters and advertisements that show women partially dressed and posed in sexually attractive ways ranging from "cute" to "seductive." Because these women are often portrayed as firefighters, such material sends the message that how a woman looks has some relevance to how she does her job. They also convey a very confusing message on sexual harassment, since many fire service men have only a vague understanding of what does or does not constitute harassment. Such material is an insult to women's professionalism, as it encourages people to judge women firefighters on the basis of physical attractiveness rather than legitimate standards of performance. Exploiting and stereotyping groups of people to appeal to or titillate other groups is degrading; the fire service must reject such behaviors as other professional organizations have done. Stereotyping of women as sexual objects should be prohibited in the same way that unacceptable ethnic portrayals are.

Examples of ways women firefighters have been sexually harassed on the job or in their volunteer fire departments:

"An officer put moves on me. I rejected him and was treated badly."

"There's nothing I can prove. Obscene things put in my bed, locker, gear. I feel more aggravated, let down, lonely, like I'll never be totally accepted."

"Comments like, 'It's much better when you work in a t-shirt. We can see your (breasts) better."

"Continuous! Posters on walls, asking for dates, requests for sex, hugging, touching, leering, urinating in front of me, being screamed at that I don't belong here..."

"I am harassed all the time. Most of the men are not well educated and have never related to a woman in other than sexual terms. They do not know how to treat a woman as an equal."

[Source: Survey of women fire fighters, Women in the Fire Service, 1990.]

Constant remarks about the physical attributes of women, and using curse words or sexual terms to refer to women (co-workers and others), are also degrading and offensive to many women workers, even if they themselves are not the object of the commentary. Such behavior, whether it is unintentional harassment or a form of "hate speech" intended to hurt the listener, serves to isolate women from their male co-workers. Striking a balance between protecting the minority group member from the intimidating effect of such speech and the speaker's First Amendment rights is an area of the law generating considerable controversy. Rules intended to prohibit and punish hate speech, from municipal regulations to college campus rules, are being challenged in a variety of legal contexts. But as the workforce changes its composition to include more women and more people of color, employers and co-workers will find that previously "acceptable" behavior and speech will not be considered acceptable within the new cultural standards of the workplace.

Gender-based "non-sexual" harassment

The EEOC definition of harassment refers to conduct "of a sexual nature." Courts have held that actions that contain no sexual element but are directed against an employee because of her or his gender are also illegal sexual harassment.[17] Illegal discriminatory conduct based on gender that interferes with women's working conditions has been excused by some employers and co-workers as an acceptable or "natural" reaction by men to the change occasioned by women entering the firehouse workplace. In the fire service, such illegal conduct has included refusal to trade shifts, train, work with or eat with the woman firefighter, as well as interference with her gear or personal property. In such situations, courts will look at the "pervasiveness" of the conduct and whether there was also harassment that was explicitly sexual. However, pervasiveness is not defined just by length of time; the offensiveness of the individual actions is also considered.[18]

> A female employee need not subject herself to an extended period of demeaning and degrading provocation before being entitled to seek theremedies provided under Title VII.[19]

Lewd language, pornographic materials, vandalism to personal property, and anonymous telephone calls can indicate a hostile work environment and be actionable under Title VII even if it is unclear who is responsible for the acts.[20] Harassing conduct that has not yet caused psychological damage but hasimpacted on anemployee's work habits and may seriously affect the psychological well-being of the employee in the future, may create an illegally abusive work environment.[21] Courts have also held that the observance of sexual harassment of others could create a hostile environment.[22]

Stopping sexual harassment: the station officer's role

All station officers or acting officers, whatever their rank, are responsible for maintaining a harassment-free work environment in their stations. Departmental procedures may assign the officer a specific role or responsibilities in the harassment complaint procedure, but even in departments that have no clear anti-harassment policy, first-line supervisors are in a crucial position with respect to preventing sexual harassment.

❑ Stop any behavior occurring in your station or crew that is clearly harassment. In order to do this, you must be able to identify harassment when it occurs and to deal with it fairly, effectively, and quietly. If the training provided by your department on these issues is inadequate, use your own initiative to get additional training and information.

❑ Educate your crew on what does and does not constitute harassment. This should be a routine part of departmental training, but if your fire department does not offer such training, or if the training has been inadequate, consider developing and implementing a comprehensive anti-harassment training program for the people you supervise. Every fire department employee should have a working knowledge of sexual harassment as well as of the department's policies and procedures.

❑ If a member of your crew brings a complaint of sexual harassment to you, be supportive. Do not downplay the reported incident or attempt to make excuses for the behavior. Follow through on the complaint swiftly, impartially, and without discussing it with those whom it does not concern. Targets of harassment will not feel free to use the system if they fear that the whole station will immediately know of their complaint. Nor is it fair to someone who has been accused of harassment for the accusation to be made known, particularly if it should turn out to be unfounded. Only those directly involved should know of the complaint; those called as witnesses should be cautioned not to discuss the matter with others.

❑ Prevent any kind of retaliation against the person who has filed the complaint. Harassment targets often believe that filing a complaint would do no good and would probably only make matters worse. If this kind of attitude prevails in your fire station, harassment will not be reported and will only escalate, creating a nightmare for the officer as well as for the target(s).

❑ Do not overreact to harassment, but do not fail to act. Discipline should follow the department's established guidelines, and should reflect whether the offense is the individual's first or just the latest in a long series; it must also be appropriate for the severity of the behavior. A firefighter guilty of attempted sexual assault would logically be disciplined more severely than one who brought copies of pornographic magazines to work.

❑ Creating a harassment-free work environment includes stopping harassment by those who are not in the workforce. Be alert to discriminatory or harassing behavior by visitors to the stations: members of the public, friends of firefighters, equipment salespeople or repairers, and others. Handling these situations will require tact and diplomacy, but any person who behaves in that way must be made aware that you do not support their behavior or attitude, and that they will not be welcome in the station if the behavior continues.

If a co-worker is being sexually harassed

Many workers who would never harass anyone themselves are guilty of tolerating harassment that occurs in their workplace. Some onlookers may mistake the target's silence for acceptance; others may not want to get involved or may fear losing the camaraderie of their co-workers by being a "spoilsport." Fair-minded and supportive co-workers, however, will refuse to condone harassing behavior in the workplace. Your support of the target, as a friend or a witness, can be crucial to her becoming aware that the behavior need not be tolerated.

❑ If you observe behavior at work that seems questionable, don't assume it's welcome just because the recipient doesn't complain. Talk with her privately, letting her know that you felt the behavior might be inappropriate. Especially if she is a new recruit or brand new in the station, remind her that she doesn't have to accept unwelcome sexual behavior, whether it involves comments, physical contact, pornographic posters or videotapes. Emphasize to her that she has your support in getting the behavior to stop.

❑ If the behavior is clearly harassment but the target is unwilling to take steps to stop it, talk with the harasser in private. Let him know that you find the behavior inappropriate and that it's not okay with you. If your own work environment is being made uncomfortable by the sexual behavior of others, you have the right to ask that the behavior stop, or even to file a sexual harassment complaint yourself.

❑ If you think your own behavior towards a co-worker, present or past, might be or might have been harassment, *stop the behavior.* To determine where to draw the line between friendly behavior or joking and harassment, consider:

- Is there an equal exchange?

-Would you do or say the same thing to your mother or father? To another person of your gender?

- Would you want to see coverage of your behavior on the six o'clock news?

Ask the co-worker about it, privately. Let him or her know that the behavior will not continue if it's a problem, and that you can still be friends. If, after you've asked about the behavior, you're still not sure about it, stop the behavior anyway. There are hundreds of different ways for women and men to interact enjoyably and professionally in the workplace; don't risk harming a positive work relationship by behaving in ways that make either of you uncomfortable.

Liability for sexual harassment

An employer may be held absolutely liable for acts of *quid pro quo* harassment by employees who function as supervisors or as surrogates for the employer. ("Absolutely liable" means that the employer is liable even if it had no actual knowledge of the acts.) The courts are divided on whether absolute liability applies to "hostile environment" harassment. The Supreme Court has rejected blanket employer liability, but has also held that absence of notice to an employer does not always insulate the employer from liability, even when the employer established an anti-discrimination policy and grievance procedure, and the employee failed to use it. The Court also found that the employer's lack of actual knowledge will not necessarily protect the employer from liability. If the harassment is sufficiently pervasive, the employer will be assumed to have "constructive" knowledge. [23]

The EEOC Guidelines emphasize that employers have an affirmative obligation to prevent sexual harassment:

Prevention is the best tool for the elimination of sexual harassment. An employer should take all steps necessary to prevent sexual harassment from occurring, such as affirmatively raising the subject, expressing strong disapproval, developing appropriate sanctions, informing employees of their right to raise and how to raise the issue of harassment under Title VII, and developing methods to sensitize all concerned. [24]

The EEOC Guidelines also state that an employer is responsible for harassing acts of its agents, regardless of whether the acts were forbidden by or known to the employer. [25] If anemployer fails toinvestigate, investigates but bungles or fails to follow its own grievance procedures, or fails to take appropriate corrective action, a fact-finder may find grounds for employer liability. Some cases have mentioned adopting a "reasonable employer" approach. Employers have also been held liable for the acts of non-supervisory employees and even of non-employees, where the employer knew, or should have known, of the conduct and failed to take immediate and appropriate corrective action. [26]

If you are being sexually harassed

Fire department employees who are the targets of sexual harassment should:

Know the department's policies prohibiting sex discrimination and sexual harassment, and the procedures for filing a complaint.

Politely and firmly tell the harasser to stop, in front of witnesses if possible. If you are unable to confront the harasser directly, write a letter and give it to the harasser in the presence of a witness. Keep a copy of the letter in a safe place, not at work. Or speak to your supervisor, to Personnel or to an EEO officer.

If the harasser repeats the conduct, inform your supervisor *immediately* and follow it up with a note or letter to the supervisor. Again, keep a copy in a safe place, not at work. It is important that you give notice of this offensive behavior to your employer and that you have a record that you did so. (Courts will generally only hold an employer liable for sexual harassment where it can be shown that the employer knew, or should have known, about the conduct.)

Document all incidents in a diary with time, place, names of witnesses, what was said or done, and an exact account of your response and any physical or emotional stress you experienced. Keep this log in a safe place, not at work.

Talk with other women on the job or who have left the job, especially those who have worked with the harasser in the past; they may also have been targets. Do not be surprised or discouraged if other women do not support your decision to fight or report the harassment.

Keep records of all positive evaluations, promotions, etc., in the event that your complaint results in retaliation against you by your employer, officer, or other firefighters.

Contact your union, labor organization or employee group for assistance. If you work under a contract, be familiar with its anti-harassment clauses and with your rights under the by-laws of your union.

Contact women's organizations for support. You are not the first woman to be victimized by this kind of behavior. Some chapters of the National Organization for Women (NOW) sponsor support groups for victims of harassment, and many chapters of 9 to 5 can give you support as well. Other women's professional and trade groups, such as Women in the Fire Service, may be able to offer advice and resources.

Be aware that sexual harassment, and the decision to take action against it, are very stressful. Take advantage of any employer-sponsored employee assistance programs. Try not to internalize guilt. Any incidents of sexual harassment that happened to you were not your fault, the harasser did not "mean well," and he was not doing it because he likes you. Sexual harassment is not something you have to tolerate because you chose to enter a non-traditional field; you do not have to adapt to the ways of your co-workers "at any cost. " If you have friends and co-workers who can offer support, including corroboration of incidents, depend on them. Be aware, however, that the general public is largely uneducated and often unsympathetic on this issue.

If informal or internal remedies fail to stop the harassment, if you lackconfidence in the internal remedies (for example, if the person in charge of investigating your complaint is the one who is harassing you, or has threatened retaliation), or if you are retaliated against for complaining about the harassment, you should consult an attorney. Seek out a lawyer who has experience in handling employment discrimination complaints. Try to obtain private counsel, even though local, state or federal human rights agencies may assign a staff attorney to handle your formal complaint. The women's bar association or working women's advocacy groups may be able to refer you to an experienced attorney.

File a formal complaint with the federal Equal Employment Opportunity Commission or with your state or local Fair Employment Practices agency. There are specific time limits for filing such complaints, depending on whether the harassment is continuing or retaliation occurred, and whether your state or locality has an FEP agency. Title VII requires that you file with the EEOC within 180 days of the last act of discrimination. In areas with state or local FEP agencies, that time period is extended by 90 days if you filed with the local agency first; some states require you to file with the local agency first. To protect yourself, file as soon as possible. You may wish to file the charge and then ask that it be held while you attempt to resolve the problem internally. You can always drop the charge if matters are resolved to your satisfaction.

The following is a generic anti-harassment policy that may suggest ideas or provide language for the development of a policy for your fire department. No such policy should be implemented without considering your department's specific needs or seeking qualified legal advice.

The _____ Fire Department is committed to providing a harassment-free work environment for all persons, regardless of their race, sex, religion, color, age, handicap, national origin or sexual orientation. Any employee who engages in such harassment, who permits employees under his/her supervision to engage in such harassment, or who retaliates or permits retaliation against an employee who reports such harassment, is guilty of misconduct and shall be subject to remedial action which may include the imposition of discipline up to and including discharge. Retaliation against witnesses or persons who participate in the investigation of harassment is also misconduct, and offenders may be subject to discipline up to and including discharge.

In the workplace, constitutionally protected speech does not include ethnic or sex-related slurs, unwelcome sexual advances, the display of derogatory graphic materials, verbal or physical conduct of a racial or sexual nature, or other forms of harassing conduct. This type of behavior is unprofessional, and produces an uncomfortable work environment. It will not be tolerated.

Some departments write the EEOC's definition of sexual harassment (see page 56) into their policy. If state law or local ordinances provide a more comprehensive definition, it may also be included. This information may be expanded by providing specific examples of unacceptable behavior such as the following.

Sexual harassment includes, but is not limited to, the following:

Unwelcome or unwanted sexual advances, including patting, pinching, brushing up against, hugging, cornering, kissing, fondling, or any other similar physical conduct considered unacceptable by another individual;

Requests or demands for sexual favors, including subtle or blatant expectations, pressures or requests for any type of sexual favor accompanied by an implied or stated promise of preferential treatment or negative consequence concerning one's employment status;

Verbal abuse or kidding that is sexually oriented and considered unacceptableby another individual, including commenting about an individual's body or appearance where such comments go beyond mere courtesy; telling "dirty jokes" that are clearly unwanted and considered offensive by others, or any sexually-oriented comments, innuendos or actions that offend others;

Engaging in any type of sexually-oriented conduct that would unreasonably interfere with another's work performance, including extending unwanted sexual attention to someone that reduces personal productivity or time available to work at assigned tasks;

Creating a work environment that is intimidating, hostile or offensive because of unwelcome or unwanted sexually-oriented conversations, suggestions, requests, demands, physical contacts or attention or display of posters, photos or calendars.

The more "entry points" there are into the complaint process, or ways a complaint can be filed, such as with the supervisor, through the City's EEO or personnel officer, the department's human relations committee, etc., the more likely it is that targets of harassment will make use of the procedure.

Complaint procedure

Individuals who experience sexual harassment should make it clear to the offending person that such behavior is offensive to them.

Upon the occurrence of an act of sexual harassment, or upon repetition of such acts, the victim should immediately report the incident to her or his supervisor, to the department's EEO counselor, or the Personnel Department's sexual harassment counselor. All employees are assured that they may make such reports without fear of retaliation or reprisal by the City, Department management, or their supervisors.

The employee has the right to speak in private with the person to whom the sexual harassment complaint is made, or to have a witness to the harassment present.

Each complaint of sexual harassment will be fully and completely investigated by the department's EEO counselor or by the Personnel Department's sexual harassment counselor. All investigations will be handled with discretion, sensitivity and due concern for the dignity of those involved, and will be as thorough as necessary. Anyone who is alleged to have committed acts of sexual harassment will be contacted during the investigation and permitted to make a statement.

All persons named as potential witnesses by the employee will be contacted as required during the course of the investigation. Any employee who has observed the incident(s) of sexual harassment should cooperate in the investigation. All employees are assured that they may cooperate in such investigation without fear of retaliation or reprisal by the City, Department management, or their supervisors.

Employees may expect a timely resolution of all complaints.

 The policy should inform employees that even if they do not use the City's complaint process, they do not lose theprotection of state and federal law. information about state FEP and federal EEO complaint procedures should be made available to all employees:

An employee may take advantage of the formal grievance procedure in resolving the complaint. An employee may also file a complaint with (or seek advice from) the City's Equal Employment Opportunity coordinator without going to a departmental superior. The employee may decline to use the City's procedures and file a complaint directly with the state Department of Human Relations or with the U.S. Equal Employment Opportunity Commission. S/he may also file with either of these agencies while simultaneously pursuing the City's complaint procedure.

Information and advice about sexual harassment may be obtained by contacting the Personnel Department's sexual harassment counselor.

Supervisors' responsibilities

All supervisorsand acting supervisors shall maintain a working environment which is free and secure from occupational hazards including sexual harassment. Any intrusion into the work location of any element which can cause an undue interference to an employee's performance of assigned duties shall not be tolerated. Supervisors should demonstrate by their own conduct that they arecommitted to providing a workenvironment free of harassment. Supervisors shall at all times refrain from harassment and retaliation and should counsel and instruct subordinates in defining and preventing harassment.

Supervisors or acting supervisors shall immediately deal with any act of sexual harassment of which they become aware. If the supervisor or acting supervisor is unable to make a resolution or feels the incident is likely to go to the grievance stage, s/he shall notify the EEO Officer within 24 hours. Every precaution shall be taken to ensure confidentiality at this informal information-gathering stage. This shall be followed by a written report.

The supervisor or acting supervisor shall immediately investigate any complaint of sexual harassment, and move to have any such incident resolved. Supervisors will notify the fire chief any time an employee complains of harassment or of conduct which could constitute harassment.

When employees report harassment by citizens, supervisors shall use all appropriate means to stop the harassing conduct. The supervisor shall promptly inform the department head or division head about the report and what was done to resolve it.

Confidentiality

All employees shall cooperate in investigating complaints of harassment. The nature of harassment violations, particularly those involving sexual harassment, requires a high degree of confidentiality and flexibility in approaches to investigation and resolution. All employees shall keep their communications in such an investigation confidential and shall disclose them only to City officials and employees who need the disclosure in order to perform their duties.

Notes:

[1] Women in the Fire Service survey, 1990; Sunset Associates/Diane Sanchez survey, 1991. These findings are consistent with sexual harassment data from women in other fields: for example, in a 1990 survey of women in the military, two-thirds said they had been sexually harassed. Overall estimates of the percentage of women in the workforce who have been sexually harassed range upwards of fifty percent. National Council for Research on Women, *Sexuul Harassment Research and Resources*, 1991, p. 9.

[2] *Ibid.*

[3] Fitzgerald, Louise F., Y. Gold, and K.F. Brock, quote in National Council for Research on Women, *Sexual Harassment Research and Resources*, 1991, p. 11.

[4] Title VII of the Civil Rights Act of 1964, as amended, 42 U.S.C. §§2000e *et seq.*

[5] *Williams v. Saxbe*, 413 F. Supp. 654 (D.D.C. 1976); 587 F.2d 1240 (D.C. Cir. 1978) *order vacated and case remanded.* 487 F. Supp. 1387 (D.D.C. 1980) [after trial *de novo* sexual harassment determined to exist]

[6] 29 C.F.R. §1604.11.

[7] *Meritor Savings Bank v. Vinson*, 106 S. Ct. 2399.

[8] *Id.* at 2404.

[9] *Gun v. Depro Circuit Systems*, 28 F.E.P. 639 (E.D. Mo. 1982). Also see *Swentek v. USAIR, Inc.*, 830 F.2d 552, 557 (4th Cir. 1987) finding that the appropriate inquiry is whether the complainant welcomed the particular sexual antics complained of and not whether she "was the kind of person who could not be offended by such comments and therefore welcomed them generally."

[10] The Civil Rights Act does not permit punitive damages against government employers. It also limits the amount of compensatory and punitive damages to no more than $50,000 to $300,000, depending on how many employees work for the employer. §1977 A(b)(3) of the Revised Statutes as added by Civil Rights Acts of 1991 §102. Such damages in cases of harassment on racial or ethnic grounds are not limited; in 1992, legislation was in Congress that would remove the caps on damages in sexual harassment cases. Compensatory damages are available for "future pecuniary losses, emotional pain, suffering, inconvenience, mental anguish, loss of enjoyment of life, and other non-pecuniary losses." *Id.*

[11] *Jenson v. Eveleth Taconite Co.*, 139 F.R.D. 657 (D.C. Mn. 1991).

[12] *Rabidue v. Osceola Refining Co.*, 805 F.2d 611 (6th Cir. 1986), *cert. denied*, 107 S. Ct. 1983 (1987).

[13] See also *Brooms v. Regal Tube Co,*. 881 F.2d 4 12 (1989), where the Seventh Circuit aligns itself with the Sixth Circuit analysis that both objective and subjective standards (a "dual-pronged" test) are appropriate in determining whether there was a hostile work environment.

[14] Goleman, D., "Sexual Harassment: It's About Power, Not Lust," *New York* Times, October 22, 1991; C1 , C12.

[15] *Ellison v. Brady*, 924 F.2d 872 (9th Cir. 1991), *Robinson v. Jacksonville Shipyards*, 54 F.E.P. 83 (M.D. Fla. 1988); *Jenson v. Eveleth Taconite Co.*, 139 F.R.D. 657 (D.Ct. Mn. 1991); *Barbetta v. Chemlan Services*, 669 F. Supp. 569 (W.D.N.Y. 1987).

[16] 669 F. Supp. at 573

[17] See McKinney v. Dole, 765 F.2d 1129, 1138 (D.C. Cir. 1985); *Hall v. Gus Construction Co.*, 842 F.2d 1010, 1013 (8th Cir. 1988); *Hicks v. Gates Rubber Co.*, 833 F.2d 1406. 1415 (10th Cir. 1987); *Bell v. Crackin Good Bakers Inc.*, 777 F.2d 1497 (11th Cir. 1985).

[18] See *Carrero v. New York City Housing Auth.*, 890 F.2d 569. 578 (2d Cir. 1989).

[19] *Andrews v. City of Philadelphia*, 895 F.2d 1469, 1485 (3d Cir. 1990): "The pervasive use of derogatory and insulting terms" could be probative of a sexually hostile environment.

[20] *Id.*

[21] *Howard v. Dept. of Air Force*, 877 F 2d 952, 955 (Fed. Cir. 1989).

[22] *EEOC v. Gurnee Inn Corp.*, 48 F.E.P. 871, 879 (N.D. 111. 1988), *aff'd*, 914 F.2d 815 (7th Cir. 1990).

[23] *Meritor Savings Bunk v. Vinson*, *op cit.*

[24] 29C.F.R. §1604.11(f).

[25] 29 C.F.R. § 1604.11(c).

[26] Non-supervisory co-worker EEOC Guidelines 29 C.F.R. §1604.11 (d); cases include *Hall v. Gus Construction*, 842 F.2d 1010 (8th Cir. 1988) [incidents of harassment were so numerous that the employer was liable for failing to discover what was going on and remedy it]. According to case law, the non-supervisory harasser is not himself liable under Title VII [he may be under tort or other local law] because he is neither an employer nor an agent of an employer under Title VII. Non-employee EEOC Guidelines 29 C.F.R. §1604.11(e); cases on the issue of employer liability for harassment by non-employee(s) include *EEOC v. Sage Realty Corp.*, 507 F. Supp. 599 (S.D.N.Y. 1981) [revealing uniform led to many incidents of harassment by non-employees].

As a fire department begins to include more minority group members and women in its ranks, it undergoes change: a change from a "monocultural" workforce (in this case, white male) to a multi-cultural one. This transition can leave firefighters and officers of the dominant group feeling threatened, angry and resentful. On the other hand, it can be dealt with pro-actively to validate and channel people's reactions, and improve communication and understanding. Training on cultural diversity issues comes under many names: "Employment Equity Training," "Anti-Harassment Education," "Professionalism and Human Awareness," etc. By any name, the training is an important tool in managing a workforce that is in cultural transition.

Incumbent firefighters sometimes write cultural diversity training off as "special treatment for the women." "They didn't have any classes like that for me when I came on," is a common reaction. But hiring women doesn't just mean that some of your firefighters will be called Stephanie and Jackie instead of Steve and Jack. Women firefighters represent a change not just in the workforce, but in how many male firefighters perceive the job, themselves, and even women in general. One primary purpose of cultural diversity training in the fire service is to help employees-incumbent male firefighters in particular-through those stresses and transitions.

The litigation that can result from discriminatory practices is expensive. The reason to do cultural diversity training, however, is not to "stay out of legal trouble" or to keep your employees from getting into trouble. Managers provide cultural diversity training for themselves and their workforce because it's smart and because it promotes fairness. Disharmony and discrimination are expensive even if things never go as far as a lawsuit. They can be costly in terms of factors that can not be assigned a dollar value: poor morale, a high turnover of employees (particularly from just those groups you are trying to recruit and retain), or a discredited reputation for the department within the community. Cultural diversity training is not about "staying out of trouble" or about "coddling" women or people of color. It is about managing change aggressively and progressively.

Some managers have been slow to recognize that miscommunication between men and women in the workplace has a strong cultural component. But male culture is the dominant culture of the firehouse. The more forcefully that culture expresses itself in the work environment, the more difficult it will be for many women to find productive and happy careers there. Understanding differences between women and men as cultural differences lifts the burden from individuals; training on these issues gives all firefighters the tools to create a more harmonious workplace.

Cultural diversity training is not about "staying out of trouble..." It is about managing change aggressively and progressively.

The goals of cultural diversity training

Cultural diversity training is not designed to change people's attitudes. It focuses instead on their behavior in the workplace, and emphasizes that whatever one's beliefs, one has the obligation to behave fairly to others at work. Learning to value diversity means learning that no one, including oneself, has all the answers.

Several fire service realities breed and reinforce prejudice and narrow vision on the part of the dominant group (white men). These include:

A tradition of cultural dominance that is now being threatened by social change and systemic upheaval;

The dominant group's ignorance of or unfamiliarity with other cultures;

The dominant group's failure to perceive itself as just one culture out of many; and

Direct competition for limited rewards.

Fortunately, other factors offer room for thedevelopment of acceptance and harmony. Many of the factors that strengthen the bonds among white male firefighters can extend to minority and women firefighters as well:

A common purpose on the job;

Frequent contact with each other in the pursuit of common goals;

The sharing of intense experiences; and

Mutual interdependence in critical situations.

Good cultural diversity training will identify the forces that encourage intolerance, and find ways to employ and build upon the cohesive forces to combat the destructive ones.

The training should make it clear that it is the workforce that is multicultural, not just the women and minority men. Employees should gain an awareness of how they fit into the mosaic and how their view of "reality" influences and limits how comfortable they are with people who are different from themselves. The training should also recognize that new issues arise when racial stereotypes are added to sexual stereotypes: for example, white male stereotypes about African-American women are different from those about African-American men or about white women.

One department's approach to cultural diversity issues

The Prince George's County, Maryland, Fire Department in 1991 undertook a unique approach to supporting workforce cultural diversity. Fire Chief Steven T. Edwards recognized that while the department had been progressive with respect to hiring women and minorities, its ability to manage workforce diversity had lagged. In response, he organized a two-day departmental conference entitled "Fireforce 2000." A group of Prince George's County firefighters and line officers whose make-up replicated that of Workforce 2000* was asked to develop recommendations for improving the management of cultural diversity that could be implemented in the department over an 18-24-month period. A conference setting was chosen because it would permit greater employee participation in decision-making and a freer flow of information and ideas.

The conference included general sessions in which speakers identified the conference objectives and presented information on demographic trends in the workforce and the nature of cultural diversity issues. A licensed social worker with expertise in organizational effectiveness and gender issues assisted the participants in examining their own issues, *recognizing the* impact of stereotypes on their lives, and creating a safe atmosphere within the conference where sensitive issues could be discussed without anyone feeling attacked. A videotape on how mental frameworks limit people's thinking was also shown to help participants develop creative solutions to problems.

For parts of both days, the participants formed smaller discussion groups to identify those problems and issues most critical to the department. Following brainstorming and evaluation, the small groups developed strategies in response to the problems, priorities, an implementation timetable, and methods for measuring the success of each strategy.

After the conference concluded, representatives attended a meeting of senior management to present their ideas: a mentoring program for professional development, sensitivity training for all department members, improvements in the promotional process, etc. The group unanimously endorsed the idea of keeping "Fireforce 2000" together for ongoing meetings to monitor the developing programs and work on issues of continuing concern.

* That is, simulating the anticipated ethnic and gender composition of the labor force in the year 2000. The Hudson Institute's 1987 report, *Workforce 2000*, predicts that white men will comprise less than half the workforce by the year 2000, and that more than 80% of the net new workforce entrants will be women and minorities.

[Thanks to Chief Steven T. Edwards and Major William Goodwin of the Prince George's County Fire Department for the above information.]

Implementing cultural diversity training

Some fire departments handle anti-harassment and cultural diversity training in-house. Because hiring outside professionals can be expensive, the temptation to "just have Captain Smith take everyone through the material" often wins out. The disadvantages to saving money in this way are considerable. Most fire departments, as well as most city/county administrations, do not employ people who have expertise in this area. The individual(s) in charge of presenting the information and facilitating the discussions must be knowledgeable in cultural diversity issues and skilled in handling verbal conflicts and situations charged with strong emotions. When they are not, the training all too often consists either of simply reading through the department's anti-harassment policies and warning the attendees of the penalties for violations, or of unleashing a free-for-all discussion on sensitive and controversial issues without being able to channel the energy positively or resolve the conflicts that arise. The result can be worse than not offering the training at all.

When good cultural diversity training is made a management priority, it will also receive budgetary priority. Hiring a professional consulting firm to develop and deliver cultural diversity training for your department does cost money. It also means that you can hire trainers who have both expertise and experience, and makes it much more likely that the training will have a positive effect.

The most realistic solution is to hire a trainer who will incorporate members of your department into the design and initial delivery of the program, and then hand off subsequent training to them. This helps negate the perception that an outside trainer does not understand firehouse reality, further demonstrates management's commitment to the program, and ensures the greatest possible "fit" between your needs and the program's content. The trainer should design a program that is geared to your department. Its schedule should accommodate firefighters' shifts, the geographic distribution of employees throughout the city or district, and the demands of emergency responses. The program's design must reflect the department's current level of acceptance of diversity and address employees' primary issues and concerns. This should mean that the trainer will come into the stations and department headquarters to seek a wide range of personnel input before the exact content of the course is finalized.

Volunteer fire departments can also benefit greatly from cultural diversity training. Disharmony within volunteer ranks can very easily lead to firefighters or officers leaving the department, a loss most volunteer organizations can not afford. Providing training that will help bridge cultural gaps and other interpersonal barriers can mean the difference between a smoothly running organization and an inefficient, unreliable one. It can also increase the possibility of recruiting new members from among previously excluded groups in the community. Financial constraints on volunteer departments will call for creative solutions to make the training possible. Consultants' donated time, joint training with other departments, and grants or other funds from community sources or state fire agencies, may all be options.

For members of fire departments that do not provide cultural diversity training, other sources are often available. State fire academies may offer or be willing to develop programs that can be brought to individual departments or offered on a regional basis. Universities with governmental affairs departments will sometimes work with fire service agencies to develop and deliver programs. Many fire service conferences now offer workshops and speakers on cultural diversity issues and resources. The Commonwealth of Virginia holds an annual two-day EEO/AA symposium for the fire service. *(See inset.)*

Guidelines for cultural diversity training

Some consultants in workforce diversity deal primarily with ethnic issues. If you select trainers from outside your department, make sure they are also familiar with gender issues in male-dominated workplaces and have included those issues as an integral part of their curriculum - not simply tacked on some information about sexual harassment.

All personnel must receive the training, but education and the commitment to change start at the top. This means that cultural diversity and anti-harassment training are for the chief and senior staff as well as for the firefighters. All managerial personnel should receive specially designed training that addresses their needs and roles as supervisors. Members of the department's upper echelons should also be present at each training session for firefighters, not to supervise or to control behavior, but to listen and learn, and to demonstrate by their presence management's commitment to the program. It may be desirable to spend part of the time in smaller-group sessions where officers are not present, in

---

A state source of cultural diversity training

The Virginia Fire Services Board and Department of Fire Programs have held annual Equal Employment Opportunity/Affirmative Action symposia since 1988. More than 300 people from several states attended the 1991 Symposium.

The Symposium offers firefighters and officers an exposure to the basic concepts of cultural diversity, with the expectation that this opportunity for learning and increased awareness will have a positive impact on both attitude and behavior at work. Participants report favorably on both the content and impact of the symposia presentations: nearly 75% indicated that their attitudes towards minority and women firefighters had changed as a result of their participation in the event. (This is particularly noteworthy when considering that people who voluntarily attend such events are more likely to be relatively progressive or open-minded already.)

[Information about the Symposium can be obtained from the Virginia Fire Services Board, Department of Fire Programs, Parham/64 Building, Suite 200; 2807 Parham Road; Richmond, VA 23294; phone 804/527-4236.]

---

order to facilitate discussion, but firefighters and line officers must see that the management team is also going through the training.

Group size should be limited to 20-25 attendees per session. Larger sessions tend to become lectures and do not provide time for each person to speak at any length. The ground rules for the discussion should allow everyone to raise their concerns but make it clear that "beating up on" any one person or group will not be permitted.

Cultural diversity training is not a "one-shot deal." The department should hold follow-up sessions for all personnel on a regular basis. These should not simply repeat and regurgitate the information presented in the original training, but should build on prior training to improve understanding and communications. The trainer should devise an evaluation process that will gather feedback on the original training. Future sessions should incorporate changes based this information. Finally and importantly, training on cultural diversity issues should be included in the recruit curriculum or other basic firefighter training and in all officer training courses conducted by the department.

There are no "quick fixes" or easy answers to the issues that surround workforce diversity. When a fire department diversifies - when it goes from being all white male to being fully inclusive of people of color and women - its leadership must also enter into a commitment to fair treatment, equal opportunity, and valuing diversity. Hiring a few women in order to "stay out of court" and dropping them to sink or swim in the male culture of the firehouse is reactive behavior, not a proactive policy. The transition from a monocultural to a multicultural fire service must be managed with intelligence and sensitivity.

One battalion chief said of his department:

> The only reason this city has been successful in retaining large numbers of women firefighters is because of the individual women's ability to put up with an incredible amount of grief, and a few men's willingness to adapt and work things out.

Those who promote change within an institution often face strong reactions. Those who represent change often become targets of that reaction. The fire service manager is in a key position to ensure that women firefighters do not shoulder the entire burden of fire service cultural transition.

Fire departments, firefighters' union locals, and other fire service groups have developed a variety of methods for providing ongoing support to women and people of color on the job. Some of these are described in the following sections.

Mentoring programs. These match up an incumbent firefighter with a newer firefighter, usually a new recruit. The mentor provides personal contact, information on unofficial "rules" and behavior standards within the organization, the benefit of the mentor's experience as guidance for the younger firefighter, and if need be, a sympathetic voice or a shoulder to cry on. Programs are voluntary and though facilitated by management, operate at the individual level to provide crucial support for the firefighter early in his/her career or volunteer service.

Workforce diversity steering committees, women's issues committees, human relations committees. These committees or task forces can operate within a fire department or union local, or on a city-wide or county-wide basis. Their function is to provide information to management and/or union leadership regarding the

When a fire department... goes from being all white male to being fully inclusive of people of color and women, its leadership must also enter into a commitment to fair treatment, equal opportunity, and valuing diversity.

best available image

concerns of women and minorities on the job. *(See inset on the San Diego Women's Issues Committee.)*

Local or national organizations of minority or women firefighters. *(See Appendix 3 for a list of national groups.)* Local support networks of women firefighters are operating in many parts of the country. These can be informal groups that give a few women the opportunity to share problems and solutions over breakfast once a month, or more established organizations that hold regular meetings, put out a newsletter, and offer workshops and speakers on topics of interest. Their networking, support, and problem-solving functions can be critical to women's performance and longevity on the job. One local network of African-American women firefighters, for example, has developed a program that assists women in preparing for fire department jobs, from strength training on through the application and interview processes.

Other sources of support. Many resource that are not gender-specific can offer services or resources that will help employees work through the stresses that accompany the process of cultural transition. These include Employee Assistance Programs (EAP's), departmental training in stress management, conflict resolution, and communications skills, and training for officers or officer candidates in leadership styles and personnel management. *(See page 73 for one such program.)*

Departmental initiatives for identifying problems and solutions. Unique responses to the needs of individual fire departments are constantly being developed. *(See inset on Prince George's County "Fireforce2000" conference,page 69, for one example.)* The range of solutions is as wide as the collective creativity of the people in your fire department can make it.

One union's support of women

Since 1985, the San Diego Fire Fighters (IAFF Local 145) have sponsored and supported a Women's Issues Committee. The committee exists to allow the union to have input from women firefighters about their concerns over issues such as parental leave, recruitment, uniforms, and station facilities, While the committee was created and is funded by the local's executive board, decisions as to how its members would be selected and how the committee would operate have been left up to its chair.

The committee meets monthly at the union office and has full access to its facilities. A liaison person from the executive board and a secretary who takes the minutes also attend the meetings. The committee has been involved in projects such as the establishment of a maternity leave policy, development of a recruitment packet funded by the local, improvements in the fit of women firefighters' uniforms, and creation of a role-model program for new recruits.

The impact of the committee has been both to make the local more responsive to women's needs, and to make women more interested in being involved with the local. As a result of the committee's work and the local's support, the name of the "California State Firemen's Association" was changed in 1990 to the"California State Firefighters' Association." CSFA has also established its own women's issues committee that serves as a resource to all California fire departments.

The Women's Issues Committee of Local 145 is a creative response to gender integration concerns. Other IAFF locals have, with guidance from the International, established Human Relations Committees that serve a comparable purpose with respect to ethnic minorities as well.

[Thanks to members of Local 145 and its Women 's Issues Committee for the information presented here.]

Conflict resolution training

In 1988, the San Francisco Fire Department entered into a consent decree over its hiring practices. This consent decree also contained a requirement that the department "establish and maintain an informal mediation process to handle internal disputes." In response to that mandate, the SFFD has developed a "Peer Mediation" program that attempts to resolve conflicts over small issues before they escalate into major conflicts and charges of discrimination or harassment.

Under the program, volunteers from both uniformed and civilian personnel undergo an intensive training course in order to become peer mediators. The course is seven days long, eight hours a day, and takes place over a two-week period. Students spend the first two days learning to understand the elements of communication, the third day exploring sources of cultural conflict in the workplace, and the final four days learning and practicing the techniques and skills of mediation.

Mediators are used in both formal and informal ways. Informally, they can use their training to defuse conflicts in their station, to communicate better with co-workers or subordinates, and to manage potential conflict situations encountered during emergency responses. Many of them have also reported being able to use the techniques off the job, to deal with situations at home.

The formal process begins when two parties to a conflict agree to have their dispute mediated, and select a mediator from the department's list. The entire process is confidential and voluntary: either individual may withdraw at any point. The mediator uses a four-step process to assist the two sides to resolve the problem:

1. Each person tells the mediator his/her version of the situation. Ground rules require mutual respect (no name-calling, no interrupting) throughout the process. The mediator asks questions to clarify the issues and each person's feelings and fears about it.

2. The mediator opens communication between the two, helping each to understand how the other views the conflict and is affected by it, and then has them talk directly with each other about specific issues.

3. The mediator and the disputants discuss options for resolving the conflict, both this time and in case it arises again.

4. The two jointly agree upon a resolution.

The program was developed by a team of consultants and a fire department task force. When it was first established, ninety department members volunteered to be trained as mediators. As word of the program's effectiveness has spread, the list has continued to grow longer. A small group of the mediators has gone through an advanced course and will be providing in-service training for the department.

[For more information about SFFD's Peer Mediation program, see the May/June 1991 issue of NFPA Journal for an article by Victoria Macklin, the department's Manager of Human Resources.]

For firefighters to function optimally, their protective clothing must fit well and comfortably... A firefighter with ill-fitting protective equipment will be inefficient and more prone to injury.

Firefighters' protective clothing and equipment must provide protection from heat and smoke. Hazardous materials gear is designed to provide an even higher level of protection. That is what the fire service has come to expect: protection. But for firefighters to function optimally, their protective clothing must fit well and comfortably without causing undue heat stress or adding excessive weight. A firefighter with ill-fitting protective equipment will be inefficient and more prone to injury.

Over the years, studies have addressed the functions of fit, weight, and heat stress in various components of the firefighter's protective clothing. In Pittsburgh, for example, the fire department conducted a fit study with its SCBA facepieces. They found that 20% of their firefighters required large facepieces, and, overall, that the department needed all three sizes that the manufacturer offered. The department combined these three sizes with three sizes of nosecups, representing nine possible configurations of masks/nosecups for maximized fit. NFPA 1500 (5-3.9) requires fit testing on an annual basis and permits only members with properly fitting facepieces to function in a hazardous atmosphere.

Women in the Fire Service (WFS) conducted a national survey of women firefighters in 1990, to which more than 400 women responded. One section of the survey pertained to uniforms and protective gear. Eighty percent of the women reported that they had experienced fit problems with protective gear or uniforms at some time during their firefighting service; 45% of the career firefighters and 56% of the volunteers indicated that they still had problems with one or more items of ill-fitting gear. (See *chart below.*)

Sources of women's protective gear and uniforms

In the fall of 1991, WFS distributed a survey to gather information on sizing and NFPA compliance of protective clothing and equipment. Items covered in the survey included firefighting boots (both rubber and leather), brushfire clothing, bunker coats and pants, firefighting gloves, hazardous materials suits, helmets, proximity suits, safety shoes, SCBA facepieces, and station uniforms. More than 140 manufacturers and distributors of protective clothing, station uniforms and breathing

Percentage of women firefighters reporting improperly fitting protective gear or station uniforms		
Item	Career firefighters	Volunteer firefighters
Gloves	25%	25%
Turnout bunker coat	17%	37%
Turnout/bunker pants	12%	34%
Boots	12%	22%
Uniforms	17%	10%
Helmet	10%	16%
SCBA facepiece	10%	21%
[Source 1990 survey. Women in the Fire Service]		

apparatus were sent a survey form and a letter explaining its purpose. The information on the following pages is taken directly from the survey forms returned by the manufacturers and distributors, who are responsible for its accuracy.

WFS also informed manufacturers of an information-gathering meeting scheduled for the 1991 International Association of Fire Chiefs' Conference in Toronto. Input received at that meeting is included below. Every manufacturer of personal protective equipment that had an exhibit at the IAFC conference was visited by a researcher for this handbook and asked either to elaborate on the survey form they had filled out or to complete a survey if they had not previously done so.

Report from meeting with representatives of the protective clothing industry

Fire departments must allow firefighters to be fitted in order to ensure a correct fit. In footwear, for example, more information is needed for proper fit besides just the length of the foot. The requirement that each member of the department must be fitted should be specified in the bid, and a range of sizes - such as half sizes and varying widths - should be also specified.

Fire chiefs and others in charge of purchasing should be aware that women's sizes and patterns are available in protective gear. To fit women properly, manufacturers need to know hip, waist, inseam, and chest measurements. Rise and torso measurements are important for both women's and men's turnout/bunker and uniform pants. Measurements should be taken by a manufacturer's representative the first time the fire department deals with the company, and again if the department decides to change styles, even if staying with the same manufacturer.

Ill-fitting protective clothing is not just an inconvenience; it is a safety issue. For truly accurate fit, a fit test should be done with the garment itself. Also, it is important to be aware that sizing is subjective: what "fits" according to one person may not feel to another as though it "fits" at all.

[Participating manufacturers: Fechheimer, Globe Firefighters Suits, Lion Apparel, Servus Footwear, The Warrington Group.]

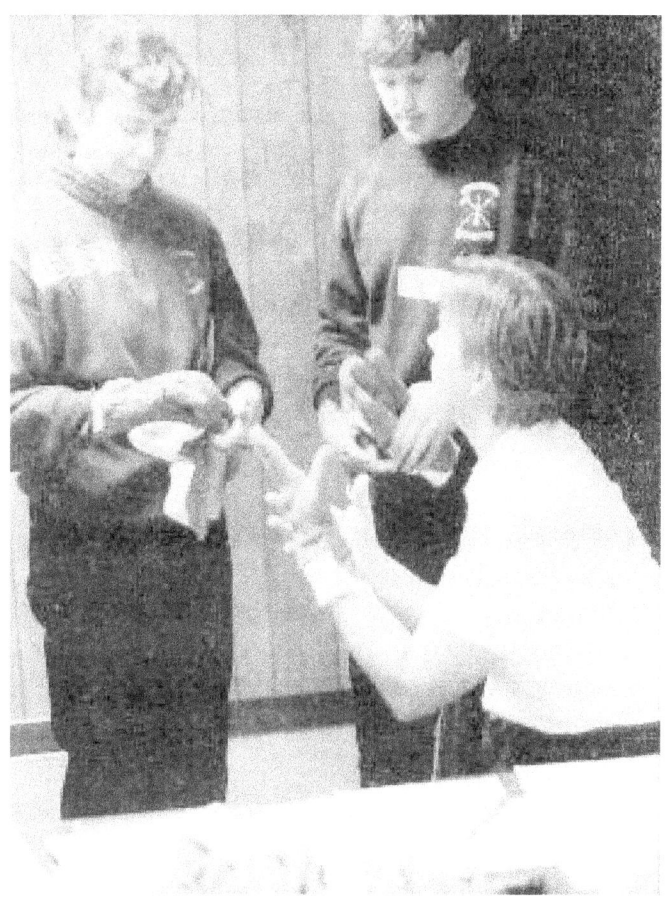

Key: The company listed is the manufacturer of the product, except as specifically noted. Clothing is made on women's patterns, and footwear on women's lasts, as noted. "Union" = union-made product. "OSHA" = meets applicable OSHA requirements. "NFPA" = meets applicable NFPA requirements. "ANSI" = meets applicable ANSI (footwear) requirements.

Rubber structural firefighting boots

Morning Pride: Women's lasts. NFPA, OSHA. Whole and half sizes, narrow and regular widths.
Neptune: (Imports) Viking #9844. Uses inserts, not women's lasts. Union, NFPA, OSHA. Sizes 5 to 15-l/2.
Ranger: #31260,31290. Women's lasts. NFPA, OSHA. Sizes 4 to 10.
Servus: Firebreaker, Firefighter. Women's lasts. Union, NFPA, OSHA. Sizes 4 to 13 narrow and wide.

Leather structural firefighting boots

Ranger: #3042,3063,3044,3046. No women's lasts. NFPA, OSHA. Sizes 5 to 13 M, 5-12 W.
Southwest: Eagle. Women's lasts. NFPA, OSHA. Sizes 3-l/2 to 15 A-EEEE.
Warrington: Warrington Pro. Unisex military last. NFPA, OSHA. Sizes 6 up.
Weinbrenner: Thorogood 804-6116. No women's lasts. OSHA.

Brush firefighting boots

Iowa American: (Distributes) Servus. Women's lasts. CAL/OSHA. Full and half sizes.

Brush firefighting jackets and pants

Alb: 06, 07, 70, 60. No women's patterns. CAL/OSHA. Sizes Small to XXL.
Safeguard America: 71-J, S-82 MOD, 71-T MOD. Women's patterns. CAL/ OSHA. Sizes Small to 6XL in jackets and shirts; pants: waist and inseam in 1" increments as required.

Structural firefighting bunker/turnout coats and pants*

Alb: C17080, 0708-0, P270080, 7008-0. No women's patterns. NFPA, OSHA. Custom sizing.
Cairns: Aegis, Traditional. Women's patterns. NFPA, OSHA. Sizes XS, S, M, L, XL, XXL.
Fire-Dex: Assault Gear, Series I, Series III. Women's patterns being developed. NFPA, OSHA. Coats sizes 32-60, pants sizes 28-56.
FireGear: [30" coat only, with pants to interface.] Women's patterns. NFPA, OSHA. Sizes 28-64 plus oversizes.
Globe: Globe, Globe Astra. Women's patterns. NFPA, OSHA. Coats sizes 28-52; pants sizes 24-52.
Iowa American: (Distributes) Quaker. Women's patterns in short-style gear. NFPA, OSHA. Custom sizing.
Lion Apparel: Body-Guard, Janesville. Women's patterns being developed. NFPA, OSHA. Coats sizes 34-60; pants sizes 30-56.
Morning Pride: [No bib-style bunker pants.] Women's patterns. NFPA, OSHA. Full range of sizes.
Ramwear: [High-waisted bunker pants rather than bib-style.] 2000, 500. Women's patterns. NFPA, OSHA. Custom sizing.

* All listed manufacturers and distributors carry bunker coats in both traditional and short styles, and bunker pants in traditional and bib styles, unless otherwise indicated.

All information presented here is provided by the individual manufacturers and distributors, who are responsible for its accuracy and completeness.

Firefighting gloves

Alb: 589. NFPA, OSHA. Sizes S, M, L, XL.
Fire-Dex: 500,600,700,800,900 series. NFPA, OSHA. Sizes XS to Jumbo.
Globe: GL-9, GL-5. NFPA (GL-9), OSHA (GL-5). Sizes XS, S, M, L, XL.
Glove Corporation: Fireman I, V, VI, VII, VIII. NFPA, OSHA. Sizes XS, S, M, L, XL, XXL.
Knoxville Glove: Fire Guardian, Fire Knox. Union, NFPA, OSHA. Sizes XS, S, M, L, XL, J.
Lion Apparel: Lion G1002, G1004, G1006, G1008. NFPA, OSHA. Sizes S, M, L, XL.
Morning Pride: NFPA, OSHA. Sizes XS, S, M, L, XL, XXL, J, and made to hand tracings.
Neptune: North Star 2990,2991,2993,2994. OSHA. Sizes XS, S, M, L, XL.
Shelby: Firewall 5221,5225,5009,5011. Union, NFPA, OSHA. Sizes XS, S, M, L, XL, J.
Tempo: Tempo Max, Tempo Pro. Union, NFPA, OSHA. Sizes XS, S, M, L, XL, J.

Hazardous materials suits

Abanda: [Level C] Fortress 7100. NFPA, OSHA. Sizes S to 7XL.
Chemfab: [Level A, Level B] Challenge 6000, Challenge 5000. Union, NFPA, OSHA. Level A: one size; Level B: two sizes.
ILC Dover: [Level A] Chemturion. NFPA (Limited use), OSHA. Sizes M, L, XL.
Interspiro: [Level A, Level B, Level C] Trellchem HI'S; Trellchem TS, TB, TL; Splash 100,200,300,700. NFPA (Level A only), OSHA. Sizes S, M, L, XL.
Kappler: [Level B, Level C] Frontline 90593. NFPA. Sizes S to 3XL.
Lion Apparel: [Level B] Pace Setter. NFPA, OSHA. Sizes XS, S, M, L, XL, XXL.

Firefighting helmets

Bullard: FH 2100, FX, Wildfire FH 911, Advent A-l. NFPA. Adjustable.
Cairns: Metro, Intruder, Patriot, Phoenix, Sam Houston, New Yorker, Philadelphian, Classic 1000, Commando, Century. NFPA, OSHA. Adjustable to size 8.
Iowa American: (Distributes) Safeco. NFPA, OSHA. Adjustable, XS to XL.
Morning Pride: Lite Force I, II, III, IV, V, VI. NFPA, OSHA. Adjustable; factory assistance for unusual problems
Rescue Equipment NW: Pacific K3/70, RK/70, F3/1. Union, NHPA, OSHA. Adjustable, 21" to 25".

SCBA facepieces

Cairns: Pioneer. NFPA, OSHA. One size.
Draeger: Draeger, Panorama, Nova. NFPA, OSHA. One size.
Interspiro: NFPA, OSHA. One size; small available "in extreme cases."
MSA: Ultravue. NFPA, OSHA. Sizes S, M, L.
Scott: NFPA, OSHA. Sizes (for 4.5 and 2.2) Small, Large, XL; (for IIa) one size.
Survivair: NFPA, OSHA.

Safety shoes

Ranger: [Sneaker style] 1542. Women's lasts. ANSI. Sizes 5-10.
Southwest: [Chukka boots; oxfords] CM-500,0X-600. No women's lasts. ANSI. Sizes 3-1/2 to 15 A-EEEE.
Warrington: Pro 2100. Women's lasts. ANSI. Sizes 6 up.
Weinbrenner: Thorogood 504-6020, 504-6010. ANSI.

Station uniforms

Artcraft Blazer: [Dress uniforms] S/800, S/801. Women's patterns. Jackets sizes 6-24; pants sizes 4-26.

Cairns: [Coveralls] Series I, Series II, Series III. Women's patterns. NFPA, OSHA. Sizes XS to XXL.

Fechheimer: [Coveralls, jackets, pants, shirts.] Paramount. Women's patterns (except coveralls). NFPA, OSHA. Pants sizes 4-24; shirts sizes 28-48.

Iowa American: [Coveralls, jackets, pants.] Quaker. Women's patterns. Union (jackets only), NFPA (jackets only), OSHA (jackets only). Any size.

Lion Apparel: [Coveralls, jackets, pants, shirts] Stationwear; NX2000 or Flamex. Women's patterns in pants and shirts. NFPA (shirts and pants), OSHA (all except jackets). Coveralls sizes S-XXXL; women's pants sizes 6-24; women's shirts sizes 6-20.

Ramwear: [Coveralls] Custom tailored.

Safeguard America: [Coveralls, jackets, pants, shirts] M-6, 74-J, 91-T, S-83, S-82. Women's patterns. NFPA, OSHA. Sizes 32XX-short to 64XX-long; waist/inseam in 1" increments; XXS-6XL.

Topps: [Coveralls, pants, shirts] Women's patterns. NFPA, OSHA. Coveralls sizes S-XXL regular and tall; pants sizes 4-28, shirts sizes 8-22.

Werner Works: [Coveralls, jackets, pants] Pro-Tuff. Women's patterns. Sizes XS3XL; women's pants and coveralls sizes 8-22.

Workrite: [Coveralls, jackets, pants, shirts] 1106N, 1104N, 3327F, 4007F, 2404F, 2434F, 2344F. Women's patterns in all except jackets. NFPA, OSHA. Sizes 34-54; pants sizes 28-50.

Firefighting proximity suits

Cairns: Aegis Proximity, Traditional Proximity. Women's patterns. NFPA, OSHA. Sizes XS, S, M, L, XL, XXL.

Fire-Dex: Aluminized. Women's patterns being developed. NFPA, OSHA.

FireGear: Women's patterns. NFPA, OSHA. Sizes 28-64 plus oversizes.

Globe: Women's patterns. NFPA. Coats sizes 28-52; pants sizes 24-52.

Lion Apparel: Body-Guard, Janesville. No women's patterns. NFPA, OSHA. Coats sizes 34-60; pants sizes 30-56.

Abanda Protective Apparel
AmSouth Building, 4th Floor
P.O. Box 2028
Decatur, AL 35602

Alb, Inc.
366 Somerville Avenue
Boston, MA 02143
617/666-8111

Artcraft Blazers
7502 Thomas St.
Pittsburgh, PA 15208
412-242-0266

E.D. Bullard Company
Route 7, Box 596
Cynthiana, KY 41031-8822
606-234-6616

Cairns & Brother, Inc.
60 Webro Road
P.O. Box 4076
Clifton, NJ 07012
201/473-5867
800-4-CAIRNS

Chemfab/Chemical Fabrics Corporation
Daniel Webster Highway
P.O. Box 1137
Merrimack, NH 03054
603-424-9000

National Draeger, Inc.
101 Technology Dr.
P.O. Box 120
Pittsburgh, PA 15230
412-787-2207

Fechheimer Brothers
(Paramount Protective Uniforms)
4545 Malsbary Road
Cincinnati, OH 45242
513-793-5400

Fire-Dex Fire Clothing
3865 W. 150th St.
Cleveland, OH 44111
216/941-3959
800-241-6563

Fire Gear, Inc.
409 AABC Suite C
Aspen, CO 81611
303/925-2303
800-748-1566

Fyrepel
P.O. Box 518
951 Buckeye Ave.
Newark, OH 43055

Globe Firefighters Suits
Loudon Road
Pittsfield, NH 03263-0128
603/435-8323

The Glove Corporation
301 N. Harrison
Alexandria, IN 46001
317-724-4481

ILC Dover, Inc.
P.O. Box 266
Frederica, DE 19946

Interspiro, Inc.
31 Business Park Dr.
Branford, CT 06405
203/481-3899

Iowa American Firefighting Equip. Co., Inc.
P.0 Box 517 Industrial Park
Osceola IA 50213
515/342-6091
800-342-IOWA

Kappler USA
P.O. Box 218
Guntersville, AL 35976
800-633-2410

Knoxville Glove Company
P.O. Box 138
Knoxville, TN 37901-0138
800-251-9738

Lion Apparel
3401 Park Center Dr.
P.O. Box 14576
Dayton, OH 45414
513/898-1949
800-421-2926

Mine Safety Appliance Co. (MSA)
P.O. Box 426
Pittsburgh, PA 15230
412-967-3167
800-MSA-2222

Morning Pride Mfg., Inc.
1986 Home Ave.
P.O. Box 557
Dayton, OH 45417
513/263-2683

Neptune International, Ltd.
6925 216th St., SW.
Lynnwood, WA 98036
206/776-5399

Ramwear, Inc.
9302 Progress Pkwy
Mentor, OH 44060
216/639-0137
800-777-9497

Ranger Footwear Co.
1100 E. Main St.
Endicott, NY 13760
607/757-4913
800-847-1307

Rescue Equipment N.W., Inc.
P.O. Box 88357
Seattle, WA 98138
206/735-0312

Safeguard America, Inc.
P.O. Box 1649
Clanton, AL 35045
205/7550-7710

Scott Aviation
225 Erie St.
Lancaster, NY 14086
716/683-5100

Servus Footwear Co., Inc.
1136 Second St.
P.O. Box 3610
Rock Island, IL 61204
309/786-7741
800-222-2668

Shelby Specialty Gloves
P.O. Box 171814
Memphis, TN 38187-1814
901-360-89828

Horace Small
P. O. Box 1269
Nashville, TN 37202-1269

Southwest Boot Company
2545 San Fernando Road, Suite 22
Los Angeles, CA 90065
213/223-2465

Survivair
3001 S. Susan St.
Santa Ana, CA 92704
714-545-0410
800-821-7236

Tempo Glove Manufacturing, Inc.
3820 W. Wisconsin Ave.
Milwaukee, WI 53208
414-344-1100
800-558-8529

Topps Manufacturing Co.
P.O. Box 750
Rochester, IN 46975
800-348-2990
800-552-2351 (in Indiana)

The Warrington Group Ltd.
Professional Products Division
216 Lafayette Road
North Hampton, NH 03862
603/964-1501
800-662-3338

Weinbrenner Shoe Co., Inc.
108 S. Polk St.
Merrill, WI 54452
715/536-5521
800-826-0002

Werner Works, Inc.
P.O. Box 974
1931 N.W. Mulholland Dr.
Roseburg, Oregon 97470
503-672-3213
800-547-0976

Workrite Uniform Co.
P.O. Box 1192
Oxnard, CA 93032
805-483-0175
800-521-1888

Bell, Laura, "Where Does Physical Testing Leave Women?" *Management Review,* December 1987, pp. 47-50.

Bird, James W., "Training Women for the P.A.T.," *Fire Engineering,* March 1991, pp. 87-93.

Booth, Walter, "Recruiting women and minorities," *Fire Chief,* May 1987, pp. 49-53. [Survey of large fire departments regarding their recruitment strategies]

Brown, Marsha D., "Getting and Keeping Women in Nontraditional Careers," *Public Personnel Management Journal,* Winter 1981, pp. 408-411.

Chambers, Mary D., "Volunteer Fire Chiefing," *International Fire Chief,* August 1980, p. 15.

Chambers, Mary D., "Women in the Fire Service," *Western Fire Journal,* April 1981, p. 40.

Craig, Jane, and R. Jacobs, "The Effect of Working with Women on Male Attitudes toward Female Firefighters," *Basic and Applied Psychology,* March 1985, pp. 61-74.

Davis, James E., "A look at performance standards," *Fire Chief,* August 1991, p. 56.

"DCFD affirmative action hiring plan overturned," *Fire Chief,* May 1987, p. 28.

Deasy, M., "One Size (Does Not) Fit All," *Firehouse,* May 1988, pp. 33-36.

"Department of Justice consent decree," *Fire Chief,* July 1987, p. 30.

Devlin & Associates, *Employment Equity Reference Manual for Ontario Municipal Fire Departments,* prepared for the office of the Fire Marshal, Ministry of the Solicitor General, 1991.

Duffy, Richard; J. Sawicki, and A. Beer, "Project FIRES: Final Report," International Association of Fire Fighters, 1985.

Durkin, Edward D., "Recruiting and Hiring Women Firefighters," *Fire Chief,* May 1981, pp. 52-55.

Evans, D.H., "Height, Weight and Physical Agility Requirements,"Journal *of Police Science and Administration,* December 1980, pp. 414-436.

Farley, Lin, *Sexual Shakedown: The Sexual Harassment of Women on the Job,* McGraw-Hill, 1978.

Feldman, Danah, "Wildland Fire Fighting," *International Fire* Chief, August 1980, pp. 16-17.

FEMA/USFA, *Physical Fitness Coordinator's Manual for Fire Departments,* 1990.

FEMA/USFA, *Stress Management: Model Program for Maintaining Firefighter Well-Being,* 1991.

Floren, Terese M., "1990 Survey Results," *WFS Quarterly,* Winter 1990-1991, pp. 14-17.

Floren, Terese M., "Women Firefighters: The Chief's Role," *Fire Chief,* May 1981, pp. 48-51.

Floren, Terese M., "Women Firefighters Speak: A Survey of the Nation's Female Firefighters," *Fire Command,* December 1980, pp. 22-24 and January 1981, pp. 22-25.

Goldfeder, William, "Retaining and Recruiting Members," *Fire Engineering,* May 1992, pp. 10-13. [Recruitment of volunteer firefighters]

Granito, Dolores, "More Women Entering Fire Service," *Fire Engineering,* March 1978, pp. 29-30.

Hamilton, Jo Carol, "Women in the Fire Service," *Fire Chief,* August 1978, pp. 81-84.

Hammond, Ken, "Recruiting women firefighters," *Fire Chief,* October 1987, pp. 40-41. [Preparing the department and the spouses of male firefighters for the entry of women into firefighting positions]

Istvan, Sharon, "Fire Protection Engineering," *International Fire Chief,* August 1980, p. 21.

Johnston, William B., *Workforce 2000: Work and Workers for the 21st Century,* Hudson Institute, 1987.

"Justice actions," *Fire Chief,* November 1986, p. 25. [Charleston, WV, Police Department ordered to reinstate dispatcher fired when she became pregnant]

"Justice Department acts in discrimination cases," *Fire Chief,* July 1985, pp. 8-10. [Deletes numerical hiring goals implemented in 1977 consent decree in San Diego]

"Justice Department sues city," *Fire Chief,* November 1991, p. 38. [Dept. of Justice alleges discrimination against white men when West Palm Beach directed the hiring of minorities and women to fill twelve vacancies]

Kay, Herbert, "Testing recruits," *Fire Chief,* April 1989, pp. 70+.

Keene, Kathy, "What is it like to be a female firefighter?*Fire Chief,* September 1991, pp. 72-74.

Larkin, Susan R., "Training for Success," *Fire Command,* August 1989, pp. 38-42. [Training women for FDNY entry-level testing]

Lipkin, Harriett, "Smoothing the Way for Women," *International Fire Chief,* August 1980, pp. 22-24.

Loden, Marilyn, *Workforce America! Managing Employee Diversity as a Vital Resource,* Business One Irwin, 1991.

MacKinnon, Catherine A., *Sexual Harassment of Working Women,* Yale University Press, 1979.

Macklin, Victoria S., "Peer Mediation Helps Heal a House Divided," *NFPA journal,* May-June 1991, pp. 62-66.

Marinucci, Richard A., "Attracting Recruits: A Matter of Image, "FireEngineering, July 1991,p 10. [Recruiting volunteer firefighters]

Marinucci, Richard A., "Women in the Volunteer Fire Service," *Fire Engineering,* January 1991, pp. 10-12.

Martin, Molly, ed., *Hard-Hatted Women,* Seal Press, 1989.

McCarl, Robert, *The District of Columbia Fire Fighters Project: A Case Study in Occupational Folklife,* Smithsonian Institute Press, 1985.

McDiarmid, Melissa, M.D.,et al.,"ReproductiveHazardsofFirefightingIandII,"American~ournalofIndustrialMedicine, 19:433-472 (1991).

McDonald, Bernie R., "Pre-Employment Training: One Department's Program," *Fire Command,* August 1987, pp. 24-28. [Louisville Fire Department]

McNichol, J., and S. Scanlin, "Proceedings of the National Firefighter Health and Safety Forum," Congressional Fire Services Institute, 1991.

McQueen, Iris, "Sexual harassment," *Fire Chief,* August 1985, pp. 69-72.

Misner, J.E., S.A. Plowman, and R.A. Bioleau, "Performance Differences between Males and Females on Simulated Firefighting Tasks," *Journal of Occupational Medicine,* 29, (1987), pp. 801-805.

Naczi, Frances D., "Removing sexism from communications," *Fire Chief,* November 1984, pp. 45-46.

Navarre, Raymond J., "Developing a stress-reducing fire station," *Fire Chief,* February 1987, pp. 46-48.

Neeves, R., *et al.,* "Physiological and Biomechanical Changes in Fire Fighters due to Boot Design Modifications," International Association of Fire Fighters and the Federal Emergency Management Agency, 1989.

National Fire Protection Association, "NFPA 1500: Standard on Fire Department Occupational Safety and Health Program," 1987.

National Fire Protection Association, "NFPA 1971: Standard on Protective Clothing for Structural Fire Fighting," 1991.

National Fire Protection Association, "NFPA 1972: Standard on Helmets for Structural Fire Fighting," 1987.

National Fire Protection Association, "NFPA 1973: Standard on Gloves for Structural Fire Fighting," 1988.

National Fire Protection Association, "NFPA 1974: Standard on Protective Footwear for Structural Fire Fighting," 1987.

National Fire Protection Association, "NFPA 1975: Standard on Station/ Work Uniforms for Structural Fire Fighters," 1990.

National Fire Protection Association, "NFPA 1981: Standard on Open-Circuit Self-Contained Breathing Apparatus for Structural Fire Fighting," 1987.

Olshan, Andrew F., K. Teschke, and I'. Baird, "Birth Defects Among Offspring of Firemen," *American Journal of Epidemiology,* Vol. 131, No. 2, pp. 312-321.

"Oregon Volunteer Firefighter of the Year: Captain Mary Lou Fletcher," *Fire Command,* September 1988, p. 6.

Osby, Robert E., "Effective fire service affirmative action," *Fire Chief* September 1991, pp. 50-54.

Paludi, Michele A., and R.B. Barickman, *Academic and Workplace Sexual Harassment,* State University of New York, 1991.

Pantoga, Fritzie, "Women Firefighters - A Survey," *Fire Chief,* January 1977, pp. 51-54.

Perkins, Kenneth B., "Volunteer Fire Fighters in the U.S.: A Sociological Profile of America's Bravest," National Volunteer Fire Council, 1987.

Petrocellil, William, and B.K. Repa, *Sexual Harassment on the Job,* Nolo Press, 1992.

Randleman, William, "What is discrimination ?" *Fire Chief,* November 1984, p. 27.

Roche, Diane C., "Public Fire Education," *International Fire Chief,* August 1980, pp. 20-21.

Rudder, Beatrice, "Career Fire Fighting," *International Fire Chief,* August 1980, p. 16.

Rukavina, John, "Fire service, meet the ADA," *Fire Chief,* June 1992, pp. 30+. [Mistitled; actually about the Civil Rights Act of 1991]

Rukavina, John, "Seeing the future," *Fire Chief,* September 1992, pp. 22+. [Regarding Pennsylvania State University report on the pending expiration of the Age Discrimination in Employment Act exemption for firefighters]

Rule, Charles H., R. E. Osby, J.H. Steffens, and M.R. Rakestraw, "Workforce 2000," *Fire Chief,* January 1991, pp. 36-40.

Sanders, Jo Schuchat, *The Nuts and Bolts of NT0: How to Help Women Enter Non-Traditional Occupations,* Scarecrow Press, 1986.

Schmidt, Wayne W., "Background investigations," *Fire Chief,* July 1990, p. 18. [Limits on background investigations in areas such as sexual activity]

Schmidt, Wayne W., "Civil liability for wrongful discharge," *Fire Chief,* September 1989, pp. 26+.

Schmidt, Wayne W., "Nepotism and consanguinity relations," *Fire Chief,* February 1985, p. 14. [9th Circuit affirms lower court ruling that one employee may be required to transfer or quit if two employees marry]

Schmidt, Wayne W., "Physical fitness standards," *Fire Chief,* July 1989, p. 42. [Legality of employment requirements restricting body fat]

Schmidt, Wayne W., "Quotas and other discrimination remedies, "Fire *Chief,* April 1985, p. 14.

Schmidt, Wayne W., "Sexual Harassment," *Fire Chief,* October 1986, pp. 14-15. [U.S. Supreme Court decision *in Meritor v. Vinson;* also Indiana case where judge found sexual harassment but not sex discrimination]

Schmidt, Wayne W., "Sexual Harassment," *Fire Chief,* February 1987, p. 27. [Appeals court reversal of lower court decision that failed to find harassment evidence of discrimination]

Schmidt, Wayne W., "Sexual Harassment," *Fire Chief,* December 1991, p. 30. [Cases involving women police officers]

Schrader, George, "Avoid sexual harassment hassles," *Fire Chief,* June 1990, pp. 47+. [Sexual harassment not an issue of sex but of power and control; recent court cases]

Schumacher, Joe, "Affirmative Action Revisited," *Fire Chief,* March 1989, pp. 51-53.

"Sex Discrimination," *Fire and Police Personnel Reporter,* August 1984, pp. 13-15.

"Sexual Harassment," *Fire and Police Personnel Reporter,* October 1986, pp. 12-14.

Shearer, Robert W., "Can after-hours conduct be grounds for firing?" *Fire Chief,* May 1989, pp. 59-60.

Shouldis, William, "Sexual Harassment," *Fire Engineering,* September 1991, pp. 101+.

Siegel, Deborah L., *Sexual Harassment: Research and Resources,* National Council for Research on Women, 1991.

Simons, George, *Working Together: How to Become More Effective in a Multicultural Organization,* Crisp Publications, 1989.

Simons, George, and D. Weissman, *Men and Women: Partners at Work,* Crisp Publications, 1990.

Smith, Michael H., "Communications skills for a changing fire service," *Fire Chief,* September 1991, p. 82.

Sturzenacker, Gloria, "Prejudice Prevention," *Chief Fire Executive,* April-May 1986, pp. 43-51.

Swartout, Robert, "Women Fire Fighters: The Seattle Concept," *International Fire Chief,* August 1980, pp. 10-11.

Tannen, Deborah, *You Just Don't Understand: Women and Men in Conversation,* Ballantine Books, 1990.

Thaut, Stanley L., "A history of Tacoma's effort to recruit women firefighters," Fire *Chief,* September 1979,   pp. 40-43.

Thomason, Betsy, "Self-discovery: a way to deal with stress," *Fire Chief,* February 1991, p. 25.

Tokle, Gary, "1001 Considerations," *Fire Command,* April 1989, pp. 24-25.

Turner, Gary, "Butting Heads over Change," *Chief Fire Executive,* January-February 1987, pp. 27-30.

"US Court of Appeals throws out FDNY scoring system," *Fire Chief,* May 1987, pp. 14-15.

Vougioukles, Carol, and L. Buchbinder, "Fire Administration Initiates Women's Program," *International Fire ChieJ* August 1980, pp. 12-13.

Vonada, Michael, "Shadow Dancing," *Fire Chief,* December 1987, p. 50.

Waters, Michael S., "The Recruitment and Retention of Women in the Career Fire Service," *International Fire Chief,* May 1986, pp. 11-17.

Webster, Cindy, "Facing off on sexual harassment," *Fire Chief,* August 1992, pp. 72-77.

Webster, Cindy, "First National Conference of Fire Service Women," *Fire Chief,* February 1986, pp. 44-47.

Williams, Timothy, & S. Evenson, "Physically fit for duty? By whose standards?" *Fire Chief,* March 1988, pp. 43+, April 1988, pp. 58+, May 1988, pp. 55+.

Willing, Linda, "Love on the Job," *Fire Chief* August 1990, p. 92.

Winkle, William, and R. Navarre, "Females in the fire service: the process of acceptance," *Fire Chief,* April 1985, pp. 68-69. [How the Toledo Fire Department integrated women into its ranks]

"Women are Fire Fighters, too!" *Fire Command,* February 1976, pp. 16-19.

"Women in the Fire Service," *Fire Chief* May 1981, pp. 48-51.

Women's Issues Advisory Committee, *Guidelines for Integration* of *Women into the California Fire Service,* California Fire Fighter Joint Apprenticeship Program, 1990.

"Women's Training Program in Jacksonville," *Fire Chief,* February 1991, pp. 60-61.

Major provisions of the Civil Rights Act of 1991

1. For the first time under federal law, permits some compensatory and punitive damages (along with attorney's fees) in cases of intentional discrimination based on sex, religion, or disability. Punitive damages are only allowed against non-public employers; specific caps are mandated on some damages.'

2. Permits jury trials in cases filed under Title VII.

3. Clarifies "mixed motive" cases to insure that if a complainant shows an employment action is motivated by an impermissible consideration of race, color, religion, sex or national origin, the practice is unlawful even if other lawful factors motivated the action or the same action would have resulted without the discriminatory motive. [Overturns *Price Waterhouse v. Hopkins* (1989).]

4. Where the complainant has shown disparate impact,[2] places the burden of proof on the employer to show that the discriminatory practice was both job-related and justified by "business necessity."[3] [Overturns *Wards Cove v. Atonio* (employer only had to show "business justification").]

5. Includes on-the-job harassment and other post-hiring conduct under the protective umbrella (prohibiting race bias) of §1981 of the Civil Rights Act of 1866 (42 U.S.C. §1981). Also expands the section to cover employers too small to be covered under Title VII, and lengthens the limitations period for filing with no requirement to file with the EEOC. [Overturns *Patterson v. McLean Credit Union (1989).]*

6. Bars challenges to consent decrees by persons who had actual notice of a proposed judgement and reasonable opportunity to present objections, and those whose interests were adequately represented by another person. Where subsequent actions are permitted, the Act gives jurisdiction to the same court and, if possible, to the same judge as in the original case. [Overturns *Martin v. Wilks* (1989).]

7. Provides that a seniority system that intentionally discriminates may be challenged when the system is adopted, when an individual becomes subject to it, or when a person is actually injured by it. [Overturns *Lorance v. AT&T* (1989).]

8. Permits expert fees to be included as part of attorney's fees for purposes of reimbursement to plaintiffs who bring actions under §1981 and Title VII. [Reverses West *Vu. Univ. Hospitals v. Casey* (1991).]

9. Stipulates that U.S. citizens employed in foreign countries by U.S.-owned or -controlled companies are covered by Title VII and the ADA, except if compliance would violate the law of the foreign country in which the company is located. [Overturns *EEOC v. ARAMCO* (1991).]

10. Prohibits the use of different cut-off scores, or otherwise altering results of employment-related tests, on the basis of race, color, religion, sex or national origin.

Following the signing of the 1991 Civil Rights Act, the federal government took a number of actions that limit its scope. The EEOC ruled that the new law does not apply to thousands of pending cases, or to any victims who file claims based on discriminatory conduct that occurred before the Act was signed (11/21/91). This may violate the intent of Congress, which explicitly exempted cases already decided (e.g., the Wards Cove plaintiffs) from the new law. As of 1992, it was not clear how this will apply to cases where there is a continuing pattern of discrimination and if discriminatory conduct prior to November 1991 can work to insulate an employer from liability for discriminatory conduct occurring after that date.

---

[1] Legislation was pending in Congress in 1992 to remove those caps.

[2] The law does not define "disparate impact," leaving intact the Supreme Court's definition in *Wards Cove v. Atonio* (1989) (the proper comparison is between the racial composition of qualified persons in the labor market and persons holding the jobs in question). Unlawful disparate impact is also shown if complainant shows there was a less discriminatory alternative available but the employer refused to use it.

[3] Business necessity is not a defense to intentional discrimination.

Sex discrimination: general

What forms of discrimination are illegal? Discrimination based on race, sex, religion, national origin, ancestry, disability, and in some states or localities, sexual orientation, political affiliation, marital status, and arrest record. Discrimination is prohibited in all terms, benefits and conditions of employment, including hiring, firing, layoff, promotion, wages and compensation, fringe benefits, assignment, and training.

Applicable laws:

Title VII of the Civil Rights Act of 1964, as amended, 42 U.S.C. §§2000e et seq.

Pregnancy Discrimination Act of 1978 (amendings §701 of Title VII), 42 U.S.C. §2000e (K): pregnancy-related conditions are to be treated the same as other disabling conditions.

Equal Pay Act of 1963, 29 U.S.C. §206(d).

State and Local Fiscal Assistance Act of 1972: "revenue-sharing" for public safety.

Americans with Disabilities Act, P.L. 101-336.

Executive Order 11246 (1965), as amended: prohibits employment discrimination on the basis of race, color, national origin or sex in institutions or agencies with federal contracts over $10,000 (including "grants" that involve a benefit to the federal government).

Intergovernmental Personnel Act of 1970: agencies or programs of state and local governments that receive grants-in-aid from the federal government

Civil Rights Act of 1871, 42 U.S.C. §1983: creates no statutory rights in and of itself, but has been used to enforce federally protected rights from other sources (not including Title VII).

Age Discrimination in Employment Act of 1967, 29 U.S.C. §621 et seq: firefighters currently exempt from prohibitions on age limits in hiring, retirement, etc. As of 1992, this exemption was under Congressional review.

The volunteer fire service and gender-based discrimination

A woman volunteer firefighter seeking to obtain the protections of Title VII would encounter the limitation that Title VII applies only to "employees." Volunteers may have legal remedies other than Title VII, but being considered as an employee will significantly increase the volunteer's protection against sex discrimination. Courts have consistently ruled that under Title VII, volunteers are not employees. [1]

In many states, however, volunteer firefighters are considered to be employees for the purposes of anti-discrimination law. This determination may depend on many factors, including how state fair employment practices laws are worded, whether volunteers receive any monetary compensation for their services, and whether they are part of a state pension system. Even if the volunteer is not considered to be an employee, state tort actions would still be available, i.e., assault and battery, intentional infliction of emotional distress, etc. If not covered under Title VII, volunteers could still be protected by the Civil Rights Act of 1871.

---

1 *Tadros v. Colman,* 898 F.2d 10 (2d Cir. 1990). *Smith v. Berks Community Television, 657* F. Supp. 794 (E.D. Pa. 1987). The *Tadros court* said plaintiff could only be an employee if the defendant both controls his/her work and pays him/her.

2 42 U.S.C. §1983

Reproductive issues

Two U.S. Supreme Court decisions, *Johnson Controls* and *California Federal Savings & Laon,* have significant impact on maternity leave issues and the rights of fertile and/or pregnant workers.

UAW v. Johnson Controls 111 S. Ct. 1196 (1991)

The EEOC's Policy Guidance states: "As a result of the Supreme Court's decision in Johnson Controls, policies that exclude members of one sex from a workplace for the purpose of protecting fetuses cannot be justified under Title VII. Thus, if a charging party alleges that the respondent has excluded members of one sex from employment based on a fetal protection policy, and if the investigation confirms this allegation, 'cause' should be found. It does not matter whether the employer can prove that a substance to which its workers are exposed will endanger the health of a fetus. It also does not matter whether the employer can prove that it will incur a higher cost as a result of hiring women. Individuals who can perform the essential functions of a job must be considered eligible for employment, regardless of the presence of workplace hazards to fetuses." C.C.H. Empl. Prac. Guide ¶5306 (6/28/91).

The Court's decision: Sex-specific fetal protection policies are, on their face, sex discrimination under Title VII because such policies require only female employees to produce proof that they are not capable of reproducing. Policies which exclude employees on the basis of gender and childbearing capacity rather than fertility are also sex discrimination under the Pregnancy Discrimination Act. A policy which discriminates on the basis of sex can only be defended if it is a "bona fide occupational qualification" (BFOQ). BFOQ must be defined as skills and aptitudes that affect the employee's ability to do the job. "The BFOQ is not so broad that it transforms this deep social concern (the possibility of injury to future children) into an essential aspect of battery-making." "Decisions about the welfare of future children must be left to the parents who conceive, bear, support and raise them rather than to employers who hire those parents."

Maternity policies that require a woman to leave fire suppression duty at a specified point in her pregnancy will no longer be considered legal. However, a fire department administration that offers its firefighters both education regarding the potential reproductive hazards of firefighting and meaningful alternative duty while the employee is attempting to conceive a child or is pregnant, will be offering its employees the option of having a child and having a job.

California Federal Savings & Loan v. Guerra, 107 S. Ct. 683,42 F.E.P. 1073 (1987)

The Court's decision: "Congress intended the Pregnancy Discrimination Act to be a 'floor beneath which pregnancy disability benefits may not drop - not a ceiling above which they may not rise.'" (at 692)

Anti-nepotism rules

Anti-nepotism rules are usually regarded by fact-finders as "sex-neutral" in thaton *their* face they apply to both men and women equally. A complainant must present ample statistics to show disparate impact* and that the rule is not justified by business necessity. A complainant must also check to see if state and local Fair Employment Practices (FEP) laws would (1) ban discrimination on the basis of marital status, and/or (2) enable the no-spouse rule to be defended as a BFOQ.

Relevant cases:

Yuhas v. Libby-Owens-Ford, 562 F.2d 496 (7th Cir. 1977), *cert denied,* 435 U.S. 934 (1978), rev'g411 F. Supp. 77 (N.D. Ill. 1976).

Plaintiff's statistics established prima facie case of sex discrimination but court accepted employer's "business" justifications that employment of both spouses in the came capacity would hurt the morale of other employees and of the spouses themselves.

EEOC Decision No. 75-239 (1976), CCH EEOC Decisions ¶6492.

Employer did not meet business necessity burden; e.g., previous history had allowed relatives to work and there was no evidence the employer had had problems before instituting the anti-nepotism rule.

Volchahoske v. City of Grand Island, 10 E.P.D. ¶10,247 (Neb. 1975).

Ordinance prohibiting employment of both husband and wife found unconstitutional interference with "right to marry" protected under the lst, 5th, 9th and 14th Amendments [city must show compelling interest].

Sebetic v. Hagerty, 640 F. Supp. 1274 (E.D. Wis. 1986), *aff'd* 819 F.2d 1144 (7th Cir.), *cert denied* 108 S. Ct. 235 (1987).

Rule prohibited spouses of law enforcement officers from being dispatchers. Court held no-spouse policy valid based on need to avoid dangerous incidents that may occur when dispatcher's spouse is officer on call. The rule did not significantly interfere with the right to marry, and the policy was not a pretext for sex discrimination.

Espinoza v. Thorna, 580 F.2d 346 (8th Cir. 1978)

Court applied a no-spouse rule to an unmarried couple.

---

* If the rule is "no spouse" rather than "no relative," plaintiff must establish the proposition that it is most often the wife who is terminated under the "no spouse" rule.

This *memorandum was issued by a state Fair Employment Practices Agency in finding probable cause of unfair discrimination in the design and administration of a fire department entry-level physical test.*

Memorandum

A determination has been made, based on the investigation results stated below, that there is PROBABLE CAUSE to credit Charging Party's allegation of an unfair discriminatory practice by Respondent in violation of (State) Statutes...

I. The Actions and Inactions of the Respondent City Evidence an Intent to Continue a Pattern and Practice of Discrimination Against Female Firefighter Applicants.

The May, 1988 physical abilities test given by the City of _____ to test applicants for entry level fire fighter positions was defective and discriminatory in the way it was conceived, administered, and scored. This test of 136 seconds was essentially the determinant of who would become a firefighter. As discussed in the following sections, all that need be shown to prove discrimination is that there was a significant adverse impact on female applicants which was not shown to be necessary to test for essential job skills or that there was an alternative test which is as accurate as the test given that does not as adversely affect female applicants. Here, however, virtually every time an option existed on how to construct the test the option that was most likely to adversely affect female applicants or unfairly help male applicants was selected. The nature of these basic decisions demonstrated an intent by the City to continue a long-standing pattern and practice of discrimination against women in the selection of firefighters.

By constructing a test that does not measure aerobic capacity (stamina), the City created a less rigorous, less valid test than they should have. By not selecting to test for essential job skills that women have performed well on in other cities' tests, the City unnecessarily hindered women's opportunity to be successful. As have most other jurisdictions around the country, (the City) identified stamina as a critical requisite ability for firefighters. With this knowledge and with the knowledge that women typically do well on aerobic tasks, the City nevertheless chose to use a short series of test elements which individually and in combination were anaerobic in nature and thus did not test stamina (aerobic) skills.

A test which fails to adequately measure aerobic capacity and by its omission is over-sampling anaerobic capacity and strength, needlessly reduces the job-relatedness of the test while simultaneously increasing its adverse impact against females. Since the federal Uniform Guidelines on Employee Selection (UGL) require the employer to select from among comparably job related selection procedures, the least onerous alternative, the use of a *less job related* procedure which is predictably *more onerous* to females is clearly illegal.

Further evidence of intent is found in the City's use of rank-ordering of test score results. Despite clear knowledge that rank-ordering would not be valid and would unfairly prejudice women's opportunities to become firefighters, the City chose to mandate the ranking of applicant scores and, thus, discriminate against women. The federal Uniform Guidelines on Employee Selection prohibit the rank-ordering of scores where there is adverse impact on a protected class of applicants unless it is shown that higher scores predict better performance on the job. The City has known this since at least 1985 and has been told about the problem a number of times; nevertheless, it chose to mandate, by ordinance, the rank-ordering of applicants on the firefighter exam. The City was fully aware that it had no evidence that higher scores on the test mean better performance on the job, and the City was fully aware that this arbitrary ranking would exclude women from fire suppression work.

In addition, the City's choice of which test events were made "pass/fail" and which were timed and counted in the final "score" seemed intentionally designed to not give fair credit for items women were known to excel in, making the test more burdensome on women. Two test components which were not timed were the "blind maze" which required the blind-folded applicant to follow a hose line through a maze and the "ladder climb." These activities more realistically duplicate actual fire fighting skills than many of the timed activities. These were also test components in which women would be expected to do better than men... both simulate key job skills but do not emphasize upper body strength. A study from another jurisdiction known by the Respondent found that women firefighters completed a ladder climb event significantly faster than male firefighters. (Concerns about safety on the ladder climb were resolved by another city by using a safety harness.)

Exacerbating this problem is that the tasks which were timed and used to create the rank order and cut-off scores were often testing for and rewarding behavior that is detrimental to effective firefighting. There were strong rewards for upper body strength and sprinting even when responsible firefighting prohibits sprinting. In a smoke filled environment it can lead to fatigue or smoke inhalation. Although upper body strength is a characteristic that can be tested for, it is not the sole characteristic that leads to effective firefighting skills. The ability to be a weight lifter is not necessarily the ability to be a firefighter. The test's disproportionate emphasis on only these skills further indicated the intent to exclude women firefighters.

Finally, the City's assertion that it tried to facilitate the women's entry into the fire service by encourage female applicants to participate in a one-year conditioning program at the YMCA lacks credibility. Most, if not all, of the women did devote themselves to the year-long course; however, the program emphasized aerobic rather than anaerobic conditioning. Conversely, the firefighter exam emphasized anaerobic capacity. The women, in essence, wasted a year of hard work in developing an important element of job-readiness which was admittedly not measured in the ultimate test.

As one former fire chief in another city stated, "If you were going to design a test to exclude women, you'd end up with (this) test." The experience, knowledge and expertise of those who developed the physical abilities test negate a conclusion that its severe adverse effect was unexpected or unplanned.

II. *The Physical Abilities Test Had an Unnecessary Adverse Impact on Female Applicants and is Discriminatory Even in the Absence of an Intent to Discriminate.*

The City has admitted that its physical abilities test adversely affected women applicants... that the test methods are discriminatory in effect. The City has never employed a female firefighter, none was deemed to have passed its 1988 physical exam and no women will be hired for the duration of the new eligibility list. Because of this, the City has the burden of proving that its testing procedures have a "manifest relationship to the employment in question." *Dothard v. Rawlingson,* 433 U.S. 321,329 (1977).

For reasons set forth below, the City has failed to meet its burden to show that its methodology is valid and that the extreme effect it has on women is justified:

1. The EEOC's Uniform Guidelines for Employee Selection provide that "(w)here cutoff scores are used, they should normally be set so as to be reasonable and consistent with normal expectations of acceptable proficiency within the work force,..." 29 C.F.R. 1607.5(H).

It is well-established practice to "norm" a test and set a cutoff score by using a random sample of incumbent employees in the job classification. Despite recommendation from both its Personnel Department and its consultant that a random selection would provide the best sample of performance within the Department, the Respondent normed the test using only volunteers who had been previously rated "excellent" or "superior" in aerobic fitness. When the sensible proposal was made to use as part of the norming group successful female firefighters from other jurisdictions to determine the test's validity for quality female firefighters, the proposal was rejected.

Despite the Respondent's position that new firefighters should have at least a "good" aerobic fitness rating, no current firefighters at that fitness rating were used as normers to set the cutoff standard. Since two of the normers in "excellent" or "superior" aerobic condition scored lower than the cutoff time set at two minutes and 16 seconds, it is possible to speculate that a number of firefighters in "good" condition would also fail the exam.

2. Stamina (aerobic capacity as opposed to anaerobic capacity) was one of three critical requisite abilities identified in the Respondent's report concerning the 1988 physical test. According to an exercise physiologist who reviewed the test, (this) exam was too short to measure aerobic capacity. Both he and an industrial psychologist who reviewed the Respondent's test agreed that by failing to measure aerobic capacity, the City needlessly reduced the job-relatedness of the test while simultaneously increasing its adverse impact upon females.

One of the Respondent's test developers admitted that he was concerned that the tasks be sequenced to provide more aerobic involvement, more closely simulating the work at a fire scene. He also admitted, however, that the physical test was predominantly anaerobic.

3. The validity of the test, its job-relatedness, was also seriously compromised by the instructions that participants "run" through the exercises. A number of fire service witnesses stated that such running at an actual fire scene is considered a safety violation. One of the Respondent's training staff made it clear that firefighters are taught not to run and are stopped from running if they do so in training.

Test normers and participants were also instructed to "dive" through a window obstacle headfirst and do a somersault on the other side of the wall. Again, evidence is compelling that such an exercise would be extremely dangerous on the fire ground, is not a realistic approximation of actual practice and is not taught to either new or experienced firefighters.

Other individual components of the test which were not representative of common work activities are the "hose pull" using a hose roller, and the obstacle course. One fire captain also criticized the run up five stories carrying a hose bundle, stating it "is not only a safety hazard but [is] often impractical or counterproductive since resulting fatigue may impair ability once there."

4. The Respondent's "job analysis" is inadequate for purposes of validating the physical test. While the job study was presumably based upon actual observation at fire scenes, the person conducting the study did not observe either residential or commercial fire suppression work; rather, he viewed a practice exercise involving a gas fire set at an oil refinery.

In addition, nothing in the job analysis was designed to gather crucial information necessary to guide the test developers in establishing "representativeness" in the duration of the individual tasks or the sequencing of tasks.

5. Some irregularities in the administration of the test itself adversely affected women who participated. Although somewhat adjustable, the airpacks used in the test did not fit all of the women. The Respondent discovered this problem and offered pieces of foam rubber to the women. In spite of the foam, or because of it, the airpacks were "very loose and slipped and flopped around affecting balance and speed."

Another problem was differing levels of encouragement given applicants. One woman said a fire service employee shouted to her, "Hey, I wouldn't even bother taking it [the test] if I were you, I saw two women go through it this morning! One dropped out in the middle and the other did worse than seven minutes." She states that the man laughed out loud after making the remark.

6. There is no evidence that a person's score on the physical test is predictive of his/her future performance as a firefighter. Just as the cut-off score set by the City failed to distinguish those applicants who could perform job tasks from those who could not, the ranking of scores fails to distinguish those who can perform job tasks better than others. One of the Respondent's test developers testified that the ranking of the City's test *assumes* that the higher ranked applicants would better perform the responsibilities of firefighting than those applicants lower on the list; however, he admitted that no studies have been done on the individuals in the norming group to determine whether there was a correlation between their job performance and their test score. This witness ultimately admitted that to assume that a person with a score of 75 would not perform the job as well as a person with a score of 77 is "probably a weak assumption."

The use of ranking in a selection procedure when the ranking fails to identify those who can better perform the task violates the federal Uniform Guidelines for Employee Selection. According to the Uniform Guidelines, the use of ranking in a content valid selection procedure "...should measure those aspects of performance which differentiate among levels of job performance." Uniform Guidelines for Employee Selection, 29 C.F.R. 1607.14(c)(9).

The City cannot establish that ranking serves to measure aspects of job performance because the City has no job performance data. Furthermore, when the use of ranking in a selection procedure has a greater adverse impact than use of the same selection procedure on a pass/fail basis, the Uniform Guidelines require that "...the user should have sufficient evidence of validity and utility to support the use of a ranking basis." 29 C.F.R. 1607.5(G). Review of the scores on the physical test establishes that, no matter where the cut-off score is set, ranking has an adverse impact on female applicants. As noted earlier, the City's support for the use of ranking is based only on the admittedly "weak assumption" that a person with a lower score would not perform as well as a person with a higher score.

7. Because of the very small number of things tested for and the brief total time needed to be selected for an offer of employment, each of these weaknesses in the test looms large even when taken alone, but when cumulated devastate any claim of the test's validity.

III. *Modifications of the City's Own Test or Alternatives Used in Other Jurisdictions Would increase the Validity of the Selection Process and Reduce Adverse Impact.*

Even if the City could show that the Phase II elements of its test, as well as its cut-off score and ranking procedures are job-related, it may still be deemed to be in violation of the law if the evidence establishes that other selection devices without a similar discriminatory effect could also serve the employer's legitimate interests. *Dothard,* 433 U.S. at 329. By analyzing the City's test and by surveying U.S. cities with (comparable populations), the Department has found a number of ways (this) test could be modified or replaced - all to the effect of increasing job-relatedness and decreasing adverse impact.

Shortcomings found in the Respondent's testing procedure suggest a number of alternatives which would make the physical exam methodology more job-related and hence more fair to both men and women, e.g.: (i) obtain a random selection of firefighters to norm the test, including firefighters in the "good" fitness category which is the Respondent's established standard; (ii) set a realistic cut-off score which does not eliminate qualified normers or exclude applicants capable of performing at levels of acceptable proficiency; (iii) utilize a "pass/fail" rating system in the absence of a valid ranking system; (iv) review the test components and identify changes which could result in increased aerobic capacity validity and lowered adverse impact; (v) add or increase aerobic capacity measures; (vi) include qualified female normers... identify procedures or shortcomings which unnecessarily impede female applicants; (vii) develop a more complete job analysis which more accurately establishes acceptable levels of performance.

A specific concern in reviewing (this) test was that the only exercises scored "pass/fail" were the ladder climb and blind maze, both of which are important tests. Neither activity would appear to have a significantly adverse impact upon women. Timing these events would make the test even more competitive and simultaneously reduce impact on women. One study found that female firefighters out-performed males in the ladder climb.

In addition to those modifications set forth in the previous paragraphs, there are a large number of alternative selection procedures being used throughout the country to pick firefighter candidates. Among those available are "construct-based" tests. According to witnesses, the City rejected such tests out-of-hand because of a fear they would be rejected by the firefighters union. Because of this concern, the City automatically excluded many alternatives available and in use in other jurisdictions.

This Department sent out a survey instrument to fire departments in 53 American cities with (comparable) populations... Thus far (after two weeks), over 50% have responded. Despite selection procedures that vary one from another, every city responding has been able to develop selection methods which are less burdensome on female applicants than this test. It is beyond comprehension that not one alternative is suitable for use in the Respondent city.

Note:

The Administrative Law Judge who later heard this case found for the plaintiffs (the women challenging the test). His decision stated that the test had been improperly normed (and therefore the test's cut-off point was not valid), that it did not adequately represent the job of firefighting, and that the City's validity study did not justify rank ordering of candidates. The judge did not find that there had been intentional discrimination against the women applicants. He ordered the City not to hire from the resulting list, mandating it instead to develop a new test that would measure aerobic as well as anaerobic capacity, and to score the test on a pass-fail basis unless rank ordering could be specifically justified. Part of his conclusions:

> On the basis of the entire record it is concluded that the City has failed to establish that (the test) is content valid and contains a representative sample of important or frequently performed job tasks... There is evidence in this case that tasks were selected because they were tough and not because they were representative of typical, important or critical job behaviors. Although the normers who were surveyed indicated that the test was representative of their duties, the Administrative Law Judge does not believe that content validity questionnaires they completed have any significant persuasive value. The record suggests that the normers would have endorsed any test that was difficult.

Films and videotapes:

"Intent vs. Impact, "Sexual harassment prevention videotape. Available from Anderson Davis, BNA Communications Inc., 9439 Key West Ave., Rockville, MD 20850.

"The Job Interview," 11-minute videotape presenting fire departent gender stereotypes in a humorous (role-reversal) scenario. Available from Women in the Fire Service.

"Trade Secrets: Blue Collar Women Speak Out," 23-minute film on women working in the trades. Available on loan from Chicago Women in Trades, 37 S. Ashland St., Chicago IL 60607; 312/942-1444.

"Valuing Diversity," 5-part videotape series. Available from Copeland and Griggs Productions, 302 23rd Ave., San Francisco, CA 94121. Includes titles such as "Managing Differences," "Diversity at Work," "Communicating Across Cultures," etc.

"What About You?/A toi de choisir!" 19-Pminute videotape profiling six women working in non-traditional occupations, including a firefighter. Available from Women's Bureau, Labour Canada, Ottawa, Ontario KlA 0J2, Canada; 819/953-0055.

"Would You Let Someone Do This to Your Sister?" 33-minute film on sexual harassment, also available in videotape. United Auto Workers Women's Department, 8000 E. Jefferson; Detroit, MI 48214; 313/926-5212.

Books and other print resources:

Devlin and Associates, *Employment Equity Reference Manual for Ontario Municipal Fire Departments,* prepared for the Office of the Fire Marshal, Ministry of the Solicitor General, 1991.

FEMA/USFA, *Physical Fitness Coordinator's Manual for Fire Departments,* 1990.

FEMA/USFA, *Stress Management: Model Program for Maintaining Firefighter Well-Being,* 1991.

Loden, Marilyn, *Workforce America! Managing Employee Diversity as a Vital Resource,* Business One Irwin, 1991.

MacKinnon, Catherine A., *Sexual Harassment of Working Women,* Yale University Press, 1979.

Martin, Molly, ed., *Hard-Hatted Women,* Seal Press, 1989.

Petrocelli, William, and B.K. Repo, *Sexual Harassment on the Job,* Nolo Press, 1992.

Sanders, Jo Schuchat, *The Nuts and Bolts of NTO: How to Help Women Enter Non-Traditional Occupations,* Scarecrow Press, 1986.

Siegel, Deborah L., *Sexual Harassment: Research and Resources,* National Council for Research on Women, 1991. (40-page handbook)

Simons, George, *Working Together: How to Become More Effective in a Multicultural Organization,* Crisp Publications, 1989.

Simons, George, and D. Weissman, *Men and Women: Partners at Work,* Crisp Publications, 1990.

Tannen, Deborah, You *Just* Don't *Understand: Women and Men in Conversation,* Ballantine Books, 1990.

Women's Bureau, *Work and Family Resource Kit,* U.S. Department of Labor, 200 Constitution Ave. NW, Room S-331, Washington, D.C. 20210. (Single copies available free.)

Women's Issues Advisory Committee, *Guidelines for Integration of Women into the California Fire Service,* California Fire Fighter Joint Apprenticeship Program, 1990.

Organizations and their publications:

International Association of Black Professional Fire Fighters, 1025 Connecticut Ave. NW, Suite 610; Washington DC 20036; 202/296-0157. The IABPFF has national, regional and chapter committees on Black Women in the Fire Service.

International Association of Fire Fighters, 1750 New York Ave., NW, Washington, DC 20006; 202/737-8484. The IAFF makes available to its members the IAFF Manual on Human Relations, a "Hair Kit" on fire department grooming standards, the original IAFF/USFA manual, *Managing the Entry of Women in the Fire Service,* and a synopsis of information on Pregnancy and Collective Bargaining.

9 to 5, National Association of Working Women, 614 Superior Ave., NW; Cleveland OH 44113; 216/566-9308. Job Problem Hotline: 800/522-0925 (from Ohio: 216/621-9449). Resources and guidance for women on sexual harassment and other work-related concerns.

NOW Legal Defense and Education Fund, 99 Hudson St., New York NY 10013; 212/925-6635. Information on sexual harassment; guidance on anti-harassment policy development.

Women in the Fire Service, P.O. Box 5446, Madison WI 53705; 608/233-4768. WFS publishes a quarterly periodical on fire service women's issues; information packets and sample policy language on sexual harassment, maternity/reproductive issues, and other topics; and recruitment brochures and recruitment/orientation information packets aimed at women candidates for firefighter jobs. WFS also holds biennial conferences on fire service women's issues.

Women's Legal Defense Fund, 1875 Connecticut Ave. NW, Suite 710; Washington DC 20009; 202/986-2600. Information on sexual harassment; advocacy on harassment and other sex discrimination issues.

"The Changing Face of the Fire Service" -- The Symposium

In June of 1992, the U.S. Fire Administration and Women in the Fire Service convened a symposium at the National Fire Academy entitled "The Changing Face of the Fire Service." The symposium brought together fire service leaders and others from all over the U.S. to review the draft of this manual.

USFA Administrator Olin Greene and FEMA Director Wallace Stickney were present to welcome the group. Both stated their support for FEMA/USFA's recent initiatives regarding fire service women, and encouraged the participants to share their ideas about the manual.

Copies of the manual were sent to participants before the symposium. The material was divided into four subject areas, and each person was assigned to attend work sessions on two of the four areas. These work sessions generated valuable discussion and input into the various topics; many of the suggestions received were incorporated into the final version of the manual.

Symposium attendees also reviewed many of the videotapes that the manual lists as resources for fire departments in the areas of sexual harassment and cultural diversity training. The symposium adjourned following a closing session that provided summaries from the work sessions and final discussion of the manual. Feedback from the participants was universally positive about both the manual and the symposium. Many attendees recommended that the USFA update the manual on a regular basis and hold annual symposia at the National Fire Academy on fire service women's issues.

The authors of the manual sincerely thank all of the symposium attendees for contributing their time and ideas. We also thank the U.S. Fire Administration for making both the manual and the symposium possible.

Photo credits:
Pages 4-5: Women *in the* Fire *Service*
Page 7: Vic Nicastro
Page 8: Women in the Fire Service
Page 18: Women in the Fire Service
Page 21: Women in the Fire Service
Page 33: Women *in the Fire Service*
Page 39: Black & White Studios
Page 47: Women in the Fire Service
Page 52: Women in the Fire Service
Page 68: Dan Jones
Page 71: Women in the Fire Service
Page 75: Women in the Fire Service

All illustrations by Patricia Henry.

*U S Government Printing Office: 1993 -- 717-207/80981